THE GREEN GUIDE FOR
HORSE OWNERS
AND RIDERS

THE GREEN GUIDE FOR HORSE OWNERS AND RIDERS

Sustainable Practices for Horse Care,
Stable Management, Land Use, and Riding

HEATHER COOK

Storey Publishing

The mission of Storey Publishing is to serve our customers by
publishing practical information that encourages
personal independence in harmony with the environment.

Edited by Deborah Burns and Sue Ducharme
Art direction and book design by Cynthia N. McFarland
Cover design by Alethea Morrison
Text production by Jennifer Jepson Smith

Cover and interior decorative art by © Mutt Ink
Illustrations by Michael Gellatly, except for insects page 144 by Cathy Baker
Infographics by Leslie Anne Charles
Maps by Ilona Sherratt
Photograph on page 8 © iStockphoto.com, with additional photos on pages 93 and
    164 by Mars Vilaubi

Indexed by Nancy D. Wood

Printed in the United States by Versa Press
10 9 8 7 6 5 4 3 2 1

LIBRARY OF CONGRESS CATALOGING-IN-PUBLICATION DATA

Cook, Heather, 1977–
    The green guide for horse owners and riders / by Heather Cook.
        p. cm.
    Includes index.
    ISBN 978-1-60342-147-8 (pbk. : alk. paper)
    ISBN 978-1-60342-148-5 (hardcover : alk. paper)
    1. Horses—Environmental aspects. 2. Environmental
    protection—Citizen participation. I. Title.
SF285.C59 2009
636.10028'6—dc22
                                        2009007847

CONTENTS

(continued)

PREFACE

When you consider the age of our earth, it wasn't so long ago that we thought the world was flat and the sun circled around it. There may come a day in our not-so-distant future that we look back on the early years of the environmental movement and marvel at what we did not know.

As horse owners, we have a unique position that the average consumer does not: we have an intimate relationship with one of God's creatures. Much different from your average dog or cat owner who picks up pet food at the local supermarket, we know how our horse relies on this earth for his food. We see the immediate connection between the health of our ecosystem and the health of our horses. We are often the first to recognize urban creep as cities expand into previous farmland and as civilization expands into untouched regions of our planet.

I believe that we have a responsibility to ourselves, our children, and yes, even to God, to care for this planet and preserve her resources. The single most powerful natural resource on this planet is you.

ACKNOWLEDGMENTS

Many people gave generously of their time, resources, and information to bring this book and its vision to fruition.

Top of the list is my agent, Kate Epstein, who has a keen eye for detail, a definite desire to do what is right, and the heart of an encourager. And she's also a pretty darn good person. She is possibly tied with Deb Burns, my editor on this project. You have forever spoiled me for all other editors with your patience and positive, uplifting e-mails. Guess we'll just have to work together on more projects.

I encountered many industry experts who were quite liberal with their time and patiently explained many details so I knew them inside and out and could explain them to others. Most notably I tip my environmentally friendly, low-embodied-energy, and completely recyclable hat to Alayne Blickle with HorsesForCleanWater.com, Mary Ann Simonds with MysticHorse.com, and Dr. Mylon Filkins with American Trails.org. You filled my cup many times over.

To every horse owner I encountered who let me talk about their manure piles, muddy paddocks, the fuel economy on Grandpa's pickup, and incredibly interesting water rights regulations: thank you!

And to my grandfather Fredrick Mervyn Atton (former Chief Research Biologist for the Fisheries Department of the Government of Saskatchewan), who taught me that science and faith do not need to be strange bedfellows: you are missed.

I

GETTING STARTED:
The Big Picture

THE GREAT THING about taking your first big step toward being a good steward of the earth is that there are so many places to begin. You can start in any place and go in any direction and be guaranteed to do some good for the planet. Before starting to focus on the little details of being an environmentally friendly horsekeeper, take a moment to step back and consider the bigger picture.

We are all connected in immeasurable ways. A dust cloud from Africa can travel to the Gulf of Mexico; water vapor that begins in New York can become rain in Great Britain. Do not think, as you make small changes in your life today, that you are not having a bigger impact for the rest of the planet and for years to come.

THE GREEN STABLE

WELCOME TO THE NEW WAY of keeping horses! But wait, is environmentally friendly horsekeeping really a new way, or is it an old way revisited? I think it's a little of both.

Back before the wonders of technology, prefab structures, injection-molded plastics, and chemical fertilization techniques, horsekeepers were as green as they could get. They reused and recycled everything because a whole lot of work went into making even the simplest tools. Fashioning a container for water could take you most of a day if you had to build it from scratch. But today we live in a throwaway world, where 75 million plastic bottles are sent to the landfill every single day and we think nothing of tossing plastic bags into the trash.

Now, however, a new terminology has taken root in our lives. Environmental impact studies, carbon footprints, carbon credits, global warming — all of these are words that our children are learning as early as kindergarten.

So who is the green horsekeeper in this new world? A person who:

- Takes care of the land for future generations
- Considers the impact on the land today when making any decision for a home or barn
- Minimizes impact on the entire environment, including land, air, and groundwater
- Realizes that whether this impacts the local, regional, or global environment, the green horsekeeper has a personal responsibility to protect it
- Reduces chemical products used on the horse and on the land, developing alternative approaches for pest control, sanitation, and other challenges
- Relies as little as possible on manufactured energy
- Investigates and, where possible, supports and uses renewable and sustainable energy

In this book we will discuss every aspect of your barn and its environment, both inside the barn and out: the surrounding land, water, and soil. Whether you own your own property or simply rent a stall in someone else's barn, you can find ways to care for the planet so our children will be able to enjoy horses for a long time to come.

Agriculture is one of the least regulated industries in North America, especially when compared to industries such as manufacturing and transportation. Horsekeeping, a subset within the agriculture industry, is even less regulated than other subsets such as cattle ranching and crop farming. It behooves us to take the initiative and self-regulate before we have nonhorse people telling us we have to do so. In addition, it simply makes sense — horse sense, if you will.

What the GNP Doesn't Measure

Too much and too long, we seem to have surrendered community excellence and community values in the mere accumulation of material things. Our gross national product — if we should judge America by that — counts air pollution and cigarette advertising, and ambulances to clear our highways of carnage. It counts special locks for our doors and the jails for those who break them. It counts the destruction of our redwoods and the loss of our natural wonder in chaotic sprawl. It counts napalm and the cost of a nuclear warhead, and armored cars for police who fight riots in our streets. It counts Whitman's rifle and Speck's knife, and the television programs which glorify violence in order to sell toys to our children.

Yet the gross national product does not allow for the health of our children, the quality of their education, or the joy of their play. It does not include the beauty of our poetry or the strength of our marriages; the intelligence of our public debate or the integrity of our public officials. It measures neither our wit nor our courage; neither our wisdom nor our learning; neither our compassion nor our devotion to our country; it measures everything, in short, except that which makes life worthwhile. And it tells us everything about America except why we are proud that we are Americans.

— Robert F. Kennedy,
address at University of Kansas,
Lawrence, Kansas, March 18, 1968

☼ ❄ CLIMATE VARIATIONS

At the end of most chapters in this book you will see North American climate zone references to help you apply what you've learned to your specific climate zone. Each zone has unique properties; if you have lived there for a long time, you may already have an intimate knowledge of the region, its opportunities and challenges. The in-depth appendix at the end of the book lists federal, regional, and local agencies and other resources to help you assess your unique needs.

According to the *World Book Encyclopedia*, North America has the distinction of being the only continent with every kind of climate — from the humid, tropical heat of the south to the biting, dry cold of the Arctic. In the Far North of Canada, where not even trees grow on the white plains, the temperature may peak above freezing for just a short time each summer. Conversely, the southern states of the United States, which hug sea level, receive the lion's share of rainfall and may never see snow.

On average North America has four distinct seasons with moderate amounts of precipitation: it snows in the fall and winter and rains in the spring and summer. Some areas, however, have what amounts to two lengthened seasons: long, hot summers followed by mild winters, or harsh winters superseded by cool summers. North America also has the distinction of having a span of 221°F (123°C) between its record high and low. The highest temperature ever recorded in North America was 134°F (57°C) in Death Valley in 1913, and the lowest temperature was minus 87°F (minus 66°C) at Northice in Greenland in 1954.

We will focus most of our discussion, here and throughout this book, on the following zones: subarctic; humid continental; humid oceanic; highlands; semiarid; and arid.

Subarctic Zone

This hardy, tundralike area, home to mainly boreal or coniferous forests, is characterized by low temperatures and a short growing season. Flora in this area survive on less sunlight and in lower temperatures than anywhere else. Growing seasons are very short.

The subarctic zone encompasses central and western Alaska; northern Canada, including the Yukon Territory, Northwest Territories, Nunavut, Labrador, and northern parts of British Columbia, Alberta, Saskatchewan, Manitoba, Ontario, Quebec, and Newfoundland.

Humid Continental Zone

Due to the conflict between polar and tropical air masses, this temperate region has high humidity, variable weather patterns, and a large seasonal weather variance. Locations closer to the ocean experience its moderating effect on weather patterns. Deciduous forests thrive.

The humid continental zone encompasses the midwestern and northeastern parts of the United States and the southern parts of Saskatchewan, Manitoba, Ontario, and Quebec.

U.S. AND CANADIAN CLIMATIC ZONES

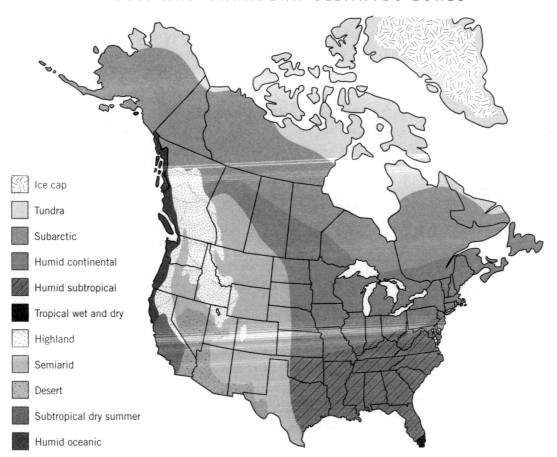

Ice cap
Tundra
Subarctic
Humid continental
Humid subtropical
Tropical wet and dry
Highland
Semiarid
Desert
Subtropical dry summer
Humid oceanic

The climate zones of North America range from tropics to tundra, from sea level to snow-capped mountains, with dozens of microclimates in between.

Humid Oceanic Zone

Found mainly along the west coast of Canada and the United States, this zone also includes the Pacific Northwest, coastal British Columbia, and the Mediterranean-type climate and chaparral biome of central and southern California. There's a narrow range of yearly temperatures across the seasons and higher precipitation amounts than in the other zones (drought is rare). This moderate fluctuation of temperature and the availability of rain and sunlight result in a wide range of flora.

The humid oceanic zone covers western Oregon; Washington; British Columbia; and northern, central, and coastal California.

Highlands Zone

This zone is characterized by higher elevations, rocky terrain, and foothills that surround mountain ranges. Most precipitation comes in the winter months in the form of snow, which melts during the summer and provides hydration during the dry months. Tree cover is sparser in higher elevations, though plentiful at the perimeter of the region due to the spring melts that provide water for trees.

The highlands zone covers the Rocky Mountains and southern parts of British Columbia.

Semiarid Zone

Consisting mainly of grasslands and shrublands that regularly experience some drought, this zone is host to many plant species that conserve water well. Higher temperatures and lower rainfall are characteristic, as is soil that can easily be depleted by overgrazing.

The semiarid zone encompasses the western parts of the Dakotas, Nebraska, Kansas, and Oklahoma; Texas; New Mexico; Colorado; Montana; Utah; and parts of Arizona, Nevada, Idaho, California, Washington, and Oregon. It also covers the areas referred to as the Great Basin, the Columbia Plateau, and the Great or Central Plains.

Arid Zone

This dry climate presents a challenge for any agricultural endeavor because the low rainfall and high temperatures dictate a reliance on mechanical forms of irrigation to support plant and animal life.

The arid zone covers "dry tropical" or desert climates found in the southwestern United States and northern Mexico.

HORSE-HUMAN ECOLOGY

How equines evolved from prey to partner,

and how that affected the environment

MANY PEOPLE ASSUME THAT horsekeeping is always green. How could anything so bucolic have a negative effect on the environment? In fact, horse people face a multitude of everyday ecological challenges: managing paddocks and pastures, controlling pests, keeping water pure and plentiful, reducing greenhouse-gas production, and so on.

When we look back over a much longer time period, tens of thousands to millions of years, the human-equine relationship becomes even more complicated. Paleontologists and biologists try to determine when our path and the horse's intertwined, diverged, and caught up to one another again. A discourse on this topic would take an entire book (and you'll find a couple of great suggestions in the bibliography), so this chapter will provide only a brief overview of how we have interacted with and managed horses throughout human history.

ROAMING THE PLAINS

According to the fossil record, horses existed in North America long before humans arrived. Scientists theorize that they roamed the plains in herds, their teeth specially developed to grind down the long, coarse prairie grasses. For millions of years the horse (or one of his ancestors) was just another prey animal — one that had many enemies and few friends, like most herbivores.

That fossil record comes to an abrupt stop, however, about 10,000 years ago, when horses inexplicably became extinct on this continent. Coincidentally, this was only a few thousand years after scientists believe humans came to North America. The exact causes of the equine extinction are unknown: was disease or drought the cause, or were human hunters at

---

DEFINITION: Horse

The word "horse" first appeared in Sumerian documents and translates into "ass of the mountains."

7

fault? Despite their disappearance from North America, horses were present on other continents by then, most notably Asia. This led scientists to theorize that horses had wandered to Asia across the Bering Land Bridge — the same bridge that brought humans to North America.

It is hard to imagine, but the first indigenous North American tribes that encountered horses hunted them for food. And why not? Horses had an abundance of meat that made a tasty meal and a hide that humans could use for clothing. If hunting was the original cause of the horse's first exit from this continent, then our initial crack at "managing" horses was a dismal failure. It wasn't until the 1600s that the horse would return, this time aboard Spanish ships.

The Wanderer

Why did this animal that had prospered so in the Colorado desert leave his amiable homeland for Siberia? There is no answer. We know that when the horse negotiated the land bridge . . . he found on the other end an opportunity for varied development that is one of the bright aspects of animal history. He wandered into France and became the mighty Percheron, and into Arabia, where he developed into a lovely poem of a horse, and into Africa where he became the brilliant zebra, and into Scotland, where he bred selectively to form the massive Clydesdale. He would also journey into Spain, where his very name would become the designation for gentleman, a *caballero*, a man of the horse. There he would flourish mightily and serve the armies that would conquer much of the known world.

— James Michener,
quoted in *Cavalry from Hoof to Track*
(see appendix B)

THE FIRST HORSEKEEPERS

By the time horses appeared again in North America, they had developed a deep relationship with humans on other continents. No one knows for sure when the horse transitioned from hunter's prey to domestic animal: some say horses were used in the Ukraine in 4000 BC, while others claim the Sumerians domesticated the horse 2,000 years later. Certainly horses were kept in confinement and used for meat or even milking long before someone climbed aboard or hooked one up to a wagon.

Nomads Become Farmers

In fact, the general transition from hunting and gathering to farming and animal domestication occurred at different times in different cultures and continents. Perhaps Native Americans would never have settled in one place at all if it had not been for the forced civilization of their land and the conquering of their people by European settlers. Some scientists believe that climate change following the last ice age (which ended about 10,000 years ago) prompted us to

Ancient art often featured the horse, from cave paintings (above) depicting him as hunters' prey to Egyptian hieroglyphics honoring him as a partner in war.

settle down as we learned to manage our food sources.

The horse's most important contribution to early civilizations was mobilizing humans, who until then had lived in relatively confined regions (going only as far as they needed to for food) and traveled in established seasonal patterns. When people began to domesticate the horse, the world opened up. Soon they could travel across continents, conquer other civilizations, and migrate to warmer climates.

Evidence of corrals dating back to between 3500 and 3000 BC has been unearthed in northern Kazakhstan. Layers of soil with high phosphorus levels (indicating manure) surrounded by postholes indicate that horses were enclosed — and that their manure wasn't always picked up!

When it came to feeding and caring for horses, humans had only to offer food from the land. Permanent stabling wasn't necessary for horses on the move, although large enclosures may have been erected to keep horses contained at summer or winter camps. Manure management and pasture rotation were not concerns for seasonally nomadic tribes, who weren't in one area long enough to make much of an impact on the land. There weren't enough humans — or horses — at the time to worry about damage to the environment, although care would have been taken to feed the horses downstream from where people pulled their water.

Areas such as the Middle East were among the first to begin farming; Eurasians had a head start of about 6,000 years on North Americans. Wherever agriculture began to take hold, farmers built barns for their horses to live in so they'd be easily accessible for farm labor, cavalry exercises (or defense), construction muscle, and transportation. Manure and urine no longer remained wherever they'd fallen but instead were picked up and spread upon the land.

A Tool of Progress

We know that shortly after horses were domesticated their main jobs were pulling and driving because ancient artwork depicts this and ancient artifacts back that up. Oxen were already pulling carts when horses were given the job. A single horse could pull much more weight than he could carry on his back, so both farmers and nomads used horses to pull implements and wagons. Riding was left to the cavalry.

And ride they did. Greeks and Romans traveled farther on horseback than on foot and thus conquered more territory. The horse became a tool, arguably the largest industrial step forward

Black Gold

An old European proverb states, "You can measure the extent of a farmer's prosperity by the height of his manure pile." In other words, the more horses a farmer had, the richer he was considered to be.

since the wheel. Huns, Mongols, Spaniards, and others developed warhorses to lead or repel invading armies. Europeans, and eventually farmers in other areas, could plant and harvest more crops using horsepower than by their hands alone and could then carry their goods and themselves to market.

RETURN TO THE WEST

By the time the Spaniards traveled to the New World, the European horse was firmly ensconced as an essential tool for both farming and conquest. Horses required incredible amounts of food during the long ocean journey, in addition to the work it took to keep them confined, healthy, and calm. If they had not been indispensable, the Spaniards would never have taken them along.

The horse — once again — revolutionized the next cultures it came into contact with: those of Native Americans. They quickly learned

Horses dramatically altered the course of history for all who came in contact with them, allowing humans to hunt more efficiently and make war more effectively.

to partner with the horse to travel farther, kill more buffalo, and go to war against other tribes. The only other "modern" invention that had an equally large culture-changing effect on Native Americans was the gun. Undoubtedly the horse in many ways was a more positive influence.

Between about 1600 and 1900 the horse was held in high regard in North America. Before his earlier extinction, he'd been hunted, but now that North America was settled, human and horse populations had increased, and a few wars had been fought upon the horse's back. For centuries nothing challenged the horse for top spot in the average person's life. Gradually, however, as cities began to develop across North America, both humans and horses became divided into two categories: urban and rural.

The Low-Impact Country Horse

Out in the countryside, horses lived difficult lives, alternating between hard, laborious farm work and long-distance travel. Although they did not have much time off, they often had some space to roam when not kept in a stall. Manure was either stored until springtime or used for fuel during winter months — not the best-smelling fire, but when living on the plains of Minnesota on a cold December evening, one didn't complain. Much.

Horses that grew up as "ranching ponies" or cow horses probably had the best lives of all — that is, the closest to the natural life of the wild horse. They had plenty of space to roam — often they were hobbled at night rather than confined to a stall. And though work could be hard, most often it involved short bursts of labor amid a lot of walking and standing. There was no manure management out on the plains; cowboys mainly treated horse manure as something to walk around. They did use both cattle and horse manure for heating fuel, however, much as indigenous people in Africa and parts of Asia had been doing for thousands of years.

The British Agricultural Revolution

The eighteenth- and nineteenth-century Agricultural Revolution, predecessor to the Industrial Revolution in Great Britain, is a prime example of how humans advanced on the backs of horses. It was characterized as a period in history with exponential growth in both agricultural productivity and output, and much of that advancement involved the horse:

- In 1730 Joseph Foljambe designs the Rotherham plough, used in Britain until the invention of the mechanized tractor
- In 1731 Jethro Tull writes *The New Horse Houghing Husbandry* about a horse-drawn seeder that would effectively seed large fields
- In 1763 John Small creates the Scots plough, which is lighter than the iron Rotherham and therefore easier for a horse team to pull
- In 1786 Andrew Meikle's threshing machine is also horse powered
- In the 1850s, when John Fowler invents a steam-powered plow, most farmers can scarcely afford it, preferring to use their horses

The Environmentally Challenged City Horse

By the nineteenth century there was a large concentration of horses inside city limits. They pulled carriages, hauled coal from mines, and transported materials and produce. And while they were far from the only environmental polluters, their impact was evident on a daily basis by the 15 to 35 pounds (7 to 16 kg) of manure and up to a gallon (4 L) of urine that each horse could produce in 24 hours. And because that horse was out working in the city, that excrement was distributed all along his route as well as in his stall.

Some growing cities tried to clean up after their horses, but this was a big task. Manure piled up along streets and out behind stables. It dried up and turned to dust, and when it rained it turned to sludge. In 1818 the New York City Council began to license "dirt carters" to gather and haul away manure.

Cartload upon cartload was stockpiled in manure yards, and men were hired to turn over the manure regularly, hardly a high-profile job and one you wouldn't have to announce to anyone, because people could probably tell by the smell of you. In 1866 the *Citizen's Association Report on the Sanitary Condition of the City* observed that "the stench arising from these accumulations of filth is intolerable."

And these large piles did more than smell; they attracted flies. Flies were so closely identified as carriers of diseases such as typhoid fever that the ones hovering over the manure piles became known as "typhoid flies." In 1908 an article appeared in *Appleton's Magazine* by a writer named Harold Bolce, who believed that most of New York City's problems came directly from the horse. He stated (with no apparent citation) that each year 20,000 New Yorkers died from "maladies that fly in the dust, created mainly by horse manure."

Life for the city horse was no picnic. The sight of dead horses in the city streets and drivers whipping their horses led to the founding of the American Society for the Prevention of Cruelty to Animals (ASPCA) in 1866. At the time a streetcar horse had a life expectancy of four years. In 1880 New York City workers removed 15,000 dead horses. But even then these horses continued to pollute because often the carcasses were simply dumped in rivers and left to rot. Eventually, rendering plants sprouted up around major cities, and horse carcasses were recycled (rendered) into gelatin, glue, and fertilizer.

Over time, as machines moved in, horses were phased out. Streetcars powered by coal also polluted the environment, but as far as city dwellers were concerned, a single-point polluter was preferable to the multipoint polluters that previously powered the vehicles.

Overloaded and overworked, city draft and carriage horses did not have a long life expectancy. During the nineteenth century their plight led to the establishment of the ASPCA in the United States and inspired Anna Sewell's beloved classic Black Beauty *in England.*

Industry In, Horses Out

After the nineteenth-century Industrial Revolution, things began to change. Horses moved out of the fast lane and no longer served to pull wagons, make deliveries, and take the family to church on Sunday. Along came cars and factories and trains and a host of other machines designed to make human lives easier and more efficient.

Only in recent decades have we realized that this gigantic industrial leap forward has had negative impacts on the earth. Pollutants and emissions make headlines on the nightly news. Every innovation that made our lives easier in the past century now seems as though it will make life harder for future generations.

FAMINE AND HARD TIMES

When the tractor replaced the horse, farmland efficiency increased greatly in North America. Farmers were able to cultivate more and better produce, and they did. But this led in turn to the overuse and destruction of large portions of farmland, and when a drought hit, crops were unable to recover.

An example of this was the Dust Bowl era from 1930 to 1936. Deep tilling and a lack of crop rotation left the topsoil of America's prairie land vulnerable to the heat and winds of a drought. The once-fertile ground simply blew away. When horses were used to plow the fields, they hadn't the power to dig as deeply as tractors did. This might have left farmers with less

Horses employed in the newly developed urban centers were often both used and abused. They were seen as a business resource, as opposed to an animal to care for, and were given the least food possible and worked until exhaustion or death.

produce but would have saved the vital topsoil from overexposure.

During the Great Depression, for a brief moment in time, horses reclaimed their status as a valued farm animal: many farmers actually returned to using horses for their farmwork because they could not afford to maintain or fuel a tractor. And yet feed was in short supply, and many horses starved to death alongside other livestock. Horses and cattle also still produced manure, of course, which could be used for fertilizing barren soil and providing heat when burned.

> " Many people have sighed for the 'good old days' and regretted the 'passing of the horse,' but today, when only those who like horses own them, it is a far better time for horses."
>
> — C. W. Anderson

LESSONS FROM HISTORY

So what have we learned about historical horse management? Several conclusions can be drawn.

First, overall, North Americans haven't always managed horses effectively. When they were prey animals, evidence suggests that we hunted them to extinction. When we began to settle down into larger cities, we spread disease by not properly managing the manure that horses produced while still managing to use the horse for our maximum benefit.

Second, horses have come in and out of favor over the years based solely on how we felt we could get them to serve us.

Third, whichever tools we use, from the horse to the newest technologies (as tractors once were), should be used responsibly and with the environment in mind so that sustainability is no longer just a keyword but a way of life. Let's learn how to do that!

As this idyllic vision shows, farm horses had a better life than the city horses did, most likely because the farmer knew the intimate connection between his work-horse and the food on his table.

II
Creating a Green Farm

WHILE ALL OF HUMANITY has the earth to care for,

you have a small piece of it that is directly under your care.

Creating a farm that integrates seamlessly into its environment,

while enhancing the natural systems, is not an easy task. This

is especially true now because we have been so conditioned to

do things in whatever way is easiest, cheapest, fastest, and most

efficient in terms of our human needs. But what of the needs of

our environment?

PUT GREEN ENERGY TO WORK

Earth-friendly solutions for your power needs

T HE TOPIC COVERED at the greatest length, depth, and breadth in green living is energy use. Often the discussion heats up when we talk about energy overuse, misuse, and abuse. Increasingly, ever since humans screwed in the first lightbulb, we've illuminated our world more brilliantly, day and night. We've powered our appliances with energy that originates at the other ends of the earth, and we frequently talk about "producing" that energy. And throughout the twentieth century, we took for granted that the energy we used was inexhaustible.

In fact, humans have never created a single energy source — we've just changed its form. Burning wood around a campfire is simply changing stored carbon into thermal energy. Making light appear in a small glass bulb can result from releasing the stored carbon in coal from the coal plant that is giving you electricity or (if you are using a greener source) from harnessing the power of the wind, sun, or water.

The Four Laws of Mother Nature

Everything is connected to everything else.
Everything has to go somewhere.
There are no free lunches.
Mother Nature always bats last.

— Author Unknown

NONRENEWABLE ENERGY

There are two types of energy sources: renewable and nonrenewable. Nonrenewable energy sources come from resources that cannot be remade, regrown, or regenerated in any way that replaces them at the rate that they're being used. These energy sources include fossil fuels (coal, petroleum, natural gas) and nuclear power. (We're going to leave the nuclear power bits to the professional scientists for now — at

THE CARBON CYCLE

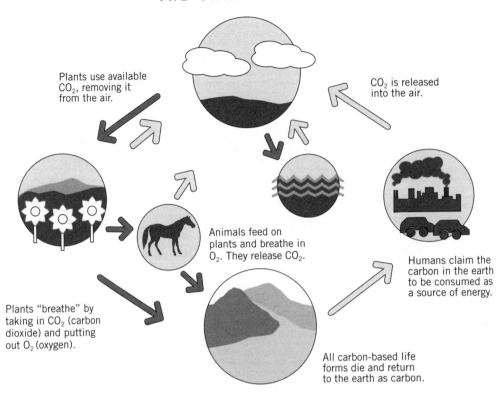

Plants use available CO$_2$, removing it from the air.

CO$_2$ is released into the air.

Animals feed on plants and breathe in O$_2$. They release CO$_2$.

Humans claim the carbon in the earth to be consumed as a source of energy.

Plants "breathe" by taking in CO$_2$ (carbon dioxide) and putting out O$_2$ (oxygen).

All carbon-based life forms die and return to the earth as carbon.

The secret to earthly health is a well-balanced carbon cycle.

least until they make safe backyard reactors.) You might say that fossil fuels can, technically, be remade, except that we'd need millions more years and many more fossils to do so.

In contrast, renewable energy taps into natural energy flows already occurring in the world: the sun shines, the wind blows, the tides ebb and flow.

Before we make the case for renewable energy sources and why we want to use them, let's discuss why we want to avoid using non-renewable energy sources. It's no secret that we are experiencing climate change or "global warming" and that the release of carbon dioxide (CO$_2$) into the atmosphere is damaging our planet. Of course, carbon dioxide is released naturally in the carbon cycle: one part of the cycle removes CO$_2$ from the environment while another puts it back.

If nothing else interfered (for example, humans), then the amounts entering and exiting the environment would be almost equal. But humans have sped up the release of carbon dioxide at a much faster rate than the natural cycle can capture it. It's part of what we call "progress": starting with the Industrial Revolution in the 1700s, humans have released increasingly more CO$_2$ through the burning of fossil fuels. According to the Environmental Protection Agency (EPA), in 2005 global levels of CO$_2$ were 35 percent higher than they were before the Industrial Revolution. As well, our average global temperature has increased by 1.2°F (0.6°C) since that time.

CO$_2$ levels in the atmosphere have increased to 377 parts per million (ppm) from the pre-nineteenth century 280-ppm level. In a 2006 report by the British government, *Avoiding*

Dangerous Climate Change, experts stated that we need to stabilize CO_2 levels at 450 ppm or below, or risk passing something called the "tipping point" — when the earth will not be able to stabilize itself.

Our Greenhouse Blanket

The atmosphere surrounding Earth is a complex mixture of gases that traps the sun's heat much the way a greenhouse does so you can grow tomatoes in inhospitable climates. Without the presence of these greenhouse gases (GHGs), the heat would escape and the average temperature of the earth would be 91°F (33°C) cooler. Life as we know it would cease to exist.

These GHGs are like a blanket, protecting us from the elements of space. As we increase the concentration of GHGs (such as CO_2) faster than they can be used up, however, it is like increasing the density of that blanket so that more heat is trapped inside. Just a ton at a time, day after day, year after year, and gradually our world gets warmer and warmer as the "blanket" gets more dense.

GHGs found in our atmosphere are (in order of relative abundance) water vapor, carbon dioxide, methane, nitrous oxide, ozone, and hydrofluorocarbons (HFCs). Of those the most significant ones are water vapor (which causes 36 to 70 percent of the greenhouse effect and absorbs a major chunk of the incoming radiation), carbon dioxide (9 to 26 percent), methane (4 to 9 percent), and ozone (3 to 7 percent).

Don't let the smaller percentages fool you, though; some gases are more potent than others, so even though they are present in

---

THE CARBON HOOFPRINT

Like other activities, keeping a horse has an impact on our environment through use of fossil fuels and production of CO_2 and other greenhouse gases. The most common measurement of our CO_2 output is in tons. Americans release an average of 20 tons (18 metric tons) of CO_2 per year. See Resources for information on calculating your carbon footprint, or CO_2 output.

CARBON vs. CARBON DIOXIDE

One of the most common errors when discussing carbon and carbon dioxide is to use the two terms as synonyms. Joseph Romm, a senior fellow at the Center for American Progress, uses a specific clarifying paragraph in his writing:

> Some people use carbon rather than carbon dioxide as a metric. The fraction of carbon in carbon dioxide is the ratio of their weights. The atomic weight of carbon is 12 atomic mass units, while the weight of carbon dioxide is 44, because it includes two oxygen atoms that each weigh 16. So, to switch from one to the other, use the formula: One ton of carbon equals 44/12 = 11/3 = 3.67 tons (3.3 metric tons) of

carbon dioxide. Thus 11 tons (10 metric tons) of carbon dioxide equals 3 tons (2.7 metric tons) of carbon, and a price of $30 per ton of carbon dioxide equals a price of $110 per ton of carbon.

| 12 atomic mass units | 16 + 12 + 16 = 44 atomic mass units |

Carbon is a part of the carbon cycle, and it is part of life. Scientists have studied the cycle extensively in relation to how we get our energy. Carbon dioxide is the by-product that is released when we use carbon.

Relative Proportions of Greenhouse Gases (GHGs) in Earth's Atmosphere

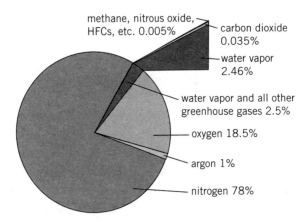

methane, nitrous oxide, HFCs, etc. 0.005%

carbon dioxide 0.035%

water vapor 2.46%

water vapor and all other greenhouse gases 2.5%

oxygen 18.5%

argon 1%

nitrogen 78%

Increase Your Sinks

A "sink" is something that removes CO_2 from the environment while a "source" puts it back in. You can increase the sinks on your property by increasing the number of plants, trees, and shrubs. During plant photosynthesis, CO_2 is removed from the atmosphere and stored as carbon in plant biomass. This is called carbon storage or "sequestration."

---

smaller concentrations, they are still critical elements. Just as a single cup (237 mL) of gasoline could poison a thousand gallons (3,785.4 L) of water, an increase of just one-half of 1 percent of methane could have a much bigger impact on our planet than a 1 percent increase in water vapor.

Carbon by itself is never a problem, it's one of the building blocks of life — everything living contains it. No matter the form, it's a helpful element. Carbon dioxide, on the other hand, is like salt in soup. Just enough CO_2 and we have a stable climate. Too much and our earth becomes too warm. To understand the use of fossil fuels, we need to realize we are "releasing" carbon dioxide when the carbon is "consumed." The carbon is never destroyed, simply converted to another form of energy, thereby releasing the by-product CO_2.

Laws and Order

To get a bit more scientific, we need to refer to the laws of thermodynamics. The first law states, "In any process, the total energy of the universe remains the same." You may recognize this as "energy cannot be created or destroyed." As

you'll see with the following alternative-energy sources, energy is simply transferred from one form into another.

The second law is a bit more complicated: "The entropy of an isolated system not in equilibrium will tend to increase over time, approaching a maximum value at equilibrium." Basically this states that as forms of energy are expended they become less easily available. That is entropy: the slow winding down of available energy. (In fact, there is a theory called "heat death," which states that eventually life in our universe will have used all available energy and there will be nothing more to use — we'll have used up all the energy we can, dissipating energy until there is none left. But don't worry; they are predicting that won't happen for 100 trillion years.)

Entropy comes into play, however, when we use regular fossil fuels like coal, oil, and gas. These highly concentrated energy sources are rapidly depleted, and when that happens it's called "high entropy." One-million-year investment . . . a few liters of oil . . . a tank of gas . . . all so you can haul your horse to a weekend horse show.

DEFINITION: **Entropy**

Entropy is a measure of the unavailability of a system's energy to do work.

RENEWABLE ENERGY

Now to get to the good stuff. Examples of renewable energy sources include water, wind, geothermal heat, and radiant energy. These are all energy forms that renew themselves naturally without assistance. We have learned to harvest this energy in different ways:

- Solar power
- Wind power
- Hydroelectricity
- Microhydro
- Biomass
- Biofuels
- Geothermal

Your first consideration when designing your property or renovating current structures is where your energy will come from and how you can efficiently use it. Energy use is the largest issue in local, regional, and even global discussions on green living and environmental issues. Utility use is the single highest source of CO_2 for horse owners, followed by transportation and the related emissions. There are many reasons why energy conservation and investment in green sources of energy will be important for your green farm.

Using energy sources that are renewable will enable you to live a sustainable life in which you are less impacted by shifting energy prices. You'll be more independent and will need to rely less on the laws of energy supply and demand. (The fuel company's trucks always seem to run behind schedule when you are nearly out of

fuel.) You will save money immediately and in the long run.

Working within your immediate environment and thinking about what you take out of it as well as put into it will ensure that you and your horses are healthier. You'll also feel better about yourself! There is nothing quite like knowing that you have lived even one day in unity with the land rather than just taking from it what you can.

And don't worry if you feel quite excited about the saving-money part; there are numerous altruistic and spiritual reasons for reducing our reliance on nonrenewable sources of energy, and they can make you feel good, too. But it's okay to be happy to save money while saving your planet.

Microgeneration

The process of harvesting natural energy sources to power your barn is called microgeneration. By utilizing microgeneration and energy conservation, most alternative-energy users can create excess energy that can be sold back to the power company and fed back into the hungry power grid that is providing energy for the surrounding community. Often, power companies award credits for the excess energy that you feed back into the grid.

The power grid is simply a network of poles and wires that connect the source of power generation with the consumers of electricity. However, since many densely populated areas have such a high energy draw, power companies have to avoid blackouts by building massive networks to shuttle energy around and to cope with high demands.

What is most interesting is that when solar energy is at its peak — during the summer months — energy grids experience their biggest demand. If we simply harnessed solar and thermal energy to power air-conditioning in most homes, we'd see huge decreases in the

> " Sunshine is a form of energy, wind and sea currents are manifestations of this energy. Do we make use of them? Oh no! We burn forests and coal, like tenants burning down our front door for heating. We live like wild settlers and not as though these resources belong to us."
>
> — Thomas Edison, 1916

money spent to power our homes, and fewer dollars would need to be spent on managing large energy grids.

But it all starts with individual microgeneration, which can be accomplished through solar, wind, geothermal, or hydro sources, depending on what fits your location best.

SOLAR ENERGY

The utilizing of solar energy is not new and is not unique to humans. Plants take solar energy and convert it into chemical energy through photosynthesis. Even for humans, using the sun's energy is actually quite an old concept. Many ancient cultures that lived in arid lands — such as the Anasazi and other Pueblo peoples of New Mexico, Utah, Arizona, and Colorado — used passive solar design to keep warm at night. They discovered that certain types of rocks would absorb the sun's heat during the day and stay warm during the night. They made their homes out of these rocks, and during the day the stones would absorb the heat without passing it through to the dwelling. At night the heat would be released into the cooling air.

Will Solar Power Work for You?

The first question to ask yourself is whether solar is a viable option for powering your barn. If your region has an abundance of sunny days, then solar energy could power both your barn and home; if your area has too many cloudy days, solar may not be a good option for your barn but might be used in smaller, less critical ways, such as using solar lights to power

 SOLAR MAPS

Check the Resources section to find online mapping resources for your area.

THE UNITED STATES

Solar maps provide monthly average daily total solar resource information on grid cells of approximately 25 miles by 25 miles (40 km by 40 km) in size. The insolation values represent the resource available to a flat plate collector, such as a photovoltaic panel, oriented due south at an angle from horizontal to equal to the latitude of the collector location. This is typical practice for PV system installation, although other orientations are also used.

CANADA

Interactive maps of the photovoltaic (PV) potential and solar resource for Canada have been developed by the Canadian Forest Service (Great Lakes Forestry Centre) in collaboration with the CANMET Energy Technology Centre (CETC-Varennes) Photovoltaic systems group. You can also gather information from your local municipality that has more specific sunlight data.

walkways or using a solar-powered temporary fence if needed. We'll discuss types of solar power from simplest to most complex.

If you live in the United States, you can check the solar rating for your area at FindSolar.com. The purpose of this site is to connect you with "solar pros" in your area, and the Frequently Asked Questions page has links to government agencies that can provide you with state-specific information on any rebate or tax-incentive programs for your property.

Passive Solar

When the Pueblo peoples lived in homes built from rock, they were using passive solar heating. It's the easiest way to use renewable energy: you aren't generating anything, and you don't need any batteries to store energy. You simply position your building or barn in the best possible location to take advantage of the heat from the sun when you need it. This is especially important for barns in the northern portions of the United States and most of Canada. If you live there, you know that if it is minus 22°F (minus 30°C), the skies are clear, and the sun is shining brightly, taking advantage of the sun to warm your barn can save you from having to turn on the furnace.

Passive solar has been used in homes for many years, and those homeowners (and realtors!) know the benefits of south-facing windows. A house can receive 30 percent of its heat from passive solar energy. But the difference between a barn and a house is the mass. The rule of thumb is that you should have a 6:9 ratio of "glass to mass" in a building. This is more easily done in homes that have walls inside, whereas most barns consist of large, open areas. Imagine a barn or a house on a weigh scale. Outer walls and the roof add mass to both buildings. Then

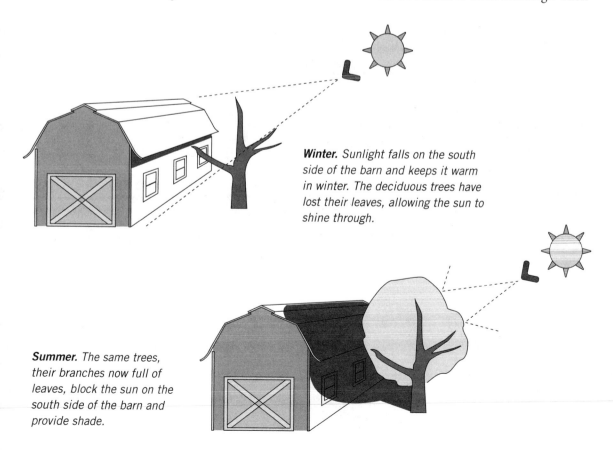

Winter. Sunlight falls on the south side of the barn and keeps it warm in winter. The deciduous trees have lost their leaves, allowing the sun to shine through.

Summer. The same trees, their branches now full of leaves, block the sun on the south side of the barn and provide shade.

in the house we add inner walls, rooms, doors, ceilings, second levels, etc. . . . So the mass of the house increases. In contrast, a barn's mass stays pretty much the same except for a few stalls.

To ensure that those large, south-facing windows are not causing elevated temperatures in the summer, you need to invest in some kind of shade. You can install outdoor shades or plant deciduous trees that provide shade in the summer and allow light through in the winter when they drop their leaves. You can also calculate how much overhang your roof should have to shade your barn during the peak hours. As long as you know your latitude, any book on passive solar heating (see Resources) will have charts to help you to determine the overhang required.

We will discuss this at greater length in future chapters on building a new barn and regreening a current structure, but the short version is that you can maximize the heat from the sun in the winter by allowing sun in the southern windows. If you can also shore up the northern side of the building (perhaps by building your arena with a shared wall, or building the barn into a hillside), it will minimize heat loss and increase the mass. The Pueblo peoples built their passive solar homes into cliff walls to stabilize the temperature inside their homes at night.

Along with passive solar, we are now using solar energy in entirely new ways. If we convert it into electricity to actively harness the sun to power our coffeepots or our horse clippers, it's called photovoltaic (PV). A secondary type is called solar thermal, which is normally used for heating.

Solar Thermal

The solar thermal method uses solar energy to generate heat. Heat from the sun is collected using solar panels that are strategically positioned to maximize solar absorption throughout the day. Inside these panels there is tubing — called solar thermal collectors — and the heat (or solar energy) is transferred to the water circulating in these tubes. To prevent the water in the tubes from freezing in the winter, a mixture of water and propylene glycol may be used.

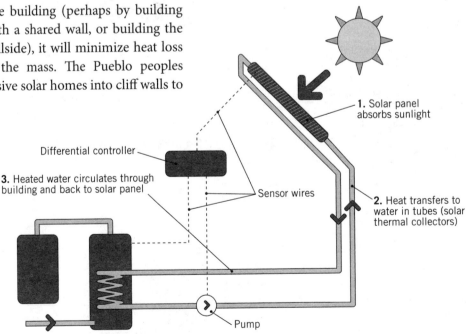

Differential controller

1. Solar panel absorbs sunlight

3. Heated water circulates through building and back to solar panel

Sensor wires

2. Heat transfers to water in tubes (solar thermal collectors)

Pump

A SOLAR THERMAL SYSTEM

Walls can be designed specifically to capture and intensify thermal energy to heat space inside a building. One example of this is a Trombe wall; the wall is made of 8- to 16-inch (20.3 to 40.6 cm)-thick masonry with a single or double layer of glass mounted on the exterior south side of the building with a 1-inch (2.5 cm) buffer between the glass and the wall. The wall must be painted a dark color to increase solar-heat absorption. The heat is stored in the wall, and when the temperature inside the structure begins to drop below the temperature of the wall, the heat radiates into the interior.

The heat will travel through the wall at about an inch (2.5 cm) per hour. So if the heat absorption begins at 1:00 P.M. and the wall is eight inches (20.3 cm) thick, then it will enter the interior space around 9:00 P.M.

Photovoltaic Cells

Often the first thought that comes to a consumer's mind when you say "green energy" is solar energy through "solar panels." These panels are actually photovoltaic (PV) cells.

In this method, a photovoltaic process converts the sun's power directly into electricity. Each solar or photovoltaic cell is made out of a silicon-based material that absorbs sunlight and often has anti-reflective material to prevent electron loss as well as glass plating to protect the cell from the elements. This differs from the previously described thermal method, which absorbs heat. Electricity (in the form of direct current, or DC) is produced when solar energy excites the electrons inside the solar cell.

Direct current (DC) is not the most accessible power because it can't move very far without losing some of its power, so it has to be transformed through an inverter into alternating current, or AC. As well, the nature of DC makes it less safe to be used in the home or barn. If you imagine the current flowing through the

A PHOTOVOLTAIC SYSTEM

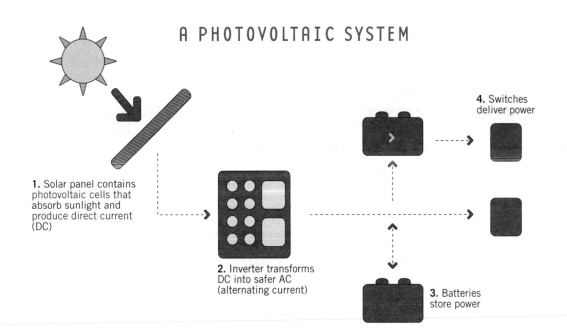

1. Solar panel contains photovoltaic cells that absorb sunlight and produce direct current (DC)

2. Inverter transforms DC into safer AC (alternating current)

3. Batteries store power

4. Switches deliver power

line like water through a hose, DC has the water flowing in one direction at all times. AC on the other hand, alternates back and forth. What this means is that if you get "zapped" by an AC current, there's a millisecond pause between the currents where you can break free. DC does not have this pause and therefore it doesn't "let go", creating the possibility of more damaging electric shocks.

The most common voltage for AC is 120V. Although a bit of energy is lost during the conversion from DC to AC, it's less than the overall loss of DC that occurs over time and distance, and the remainder is ready for use.

ON OR OFF THE GRID

You can harness photovoltaic power to run your property "off grid," meaning that all the energy you use is energy you produce through solar power. You will need a backup storage system so you don't run out of energy during cloudy, dark spells. The system is reliable when you have enough battery space available. The PV modules are connected to the battery and the battery is connected to the load. During the daytime, the batteries will remain charged and will even be able to charge when there is sun just peeking through the clouds. Nighttime is another story. During the night, your batteries will need to handle the entire load. You may want to unplug all appliances that are not required — you don't want a "phantom load" (the electricity draw coming from an appliance that's plugged in but not turned on) draining your battery.

Another option is to remain connected to the regular power grid and feed excess energy back to the grid with your PV system. Talk to your power company to determine how they manage energy-grid inputs in your area. In some states the power company is required by law to invest in green energy, and they are looking for more large-scale and microgeneration projects to meet their mandate.

WIND ENERGY

As with solar energy, use of wind energy is not a new idea. In fact, Columbus had it mastered when he "sailed the ocean blue." Five thousand years before his expeditions, Egyptians were sailing along the Nile River, and Babylonians were creating windmills in 1700 BC.

Wind results when the different surfaces of the earth (ice-capped mountains, desert valleys, dark oceans, and so on) absorb the sun's warmth at different rates. Air that's above land heats up quickly, and air over water heats up more slowly. The air that heats up first flows upward, creating a vacuum, and the cooler air rushes in to take its place. That's wind. And it's influenced by terrain, water reflectivity, time of day, season, humidity, and the rotation of the earth.

Large turbines spin when pushed by the wind and collect wind energy. A turbine is basically an alternator (like the one in your truck) that is connected to a propeller. These movements are then transformed into electrical current by electrical generators. Larger turbines at big wind farms use vertically mounted aerofoils on a rotating vertical shaft.

Wind turbines can be placed on your property (and even your roof, with special fittings to control vibration), as long as they will be 30 feet (9 m) higher than any obstruction (treetops, buildings, etc.) within 300 feet (91 m). It's important to check local building and zoning codes if installing a wind turbine sounds appropriate for you.

Need Help?

Check the Resource section for government agencies and publications that can advise you on setting up home wind systems.

TURBINE LOCATION

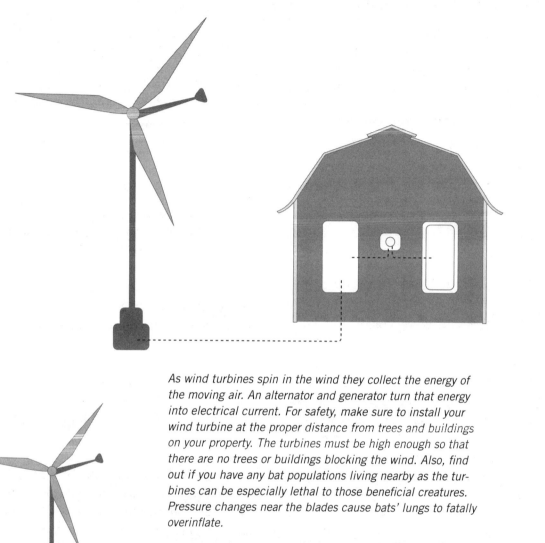

As wind turbines spin in the wind they collect the energy of the moving air. An alternator and generator turn that energy into electrical current. For safety, make sure to install your wind turbine at the proper distance from trees and buildings on your property. The turbines must be high enough so that there are no trees or buildings blocking the wind. Also, find out if you have any bat populations living nearby as the turbines can be especially lethal to those beneficial creatures. Pressure changes near the blades cause bats' lungs to fatally overinflate.

30' (9 m)

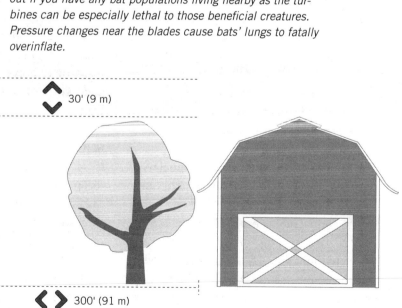

300' (91 m)

As with solar, you may be able to connect your wind turbine into the local power grid and feed clean energy back into the system. To learn how to do this, you will need to contact your local power supply company and work with them to establish the specifics about how to feed your power into their system. Most companies will have established parameters for how to draw power from a microgeneration source, and they will set this up with you.

Will Wind Power Work for You?

According to the Alternative Energy Store, there are several factors that will determine whether a wind-power system is right for you:

Conditions for stand-alone systems:

- You live in an area with average annual wind speeds of at least 9 miles (15 km) per hour (4 meters per second)
- A grid connection is not available or can only be made through an expensive power-line extension; the cost of running a power line to a remote site to connect with the utility grid can be prohibitive, ranging from $15,000 to more than $50,000 per mile (1.6 km), depending on terrain
- You have an interest in gaining energy independence from the utility
- You would like to reduce the environmental impact of electricity production
- You are aware of the intermittent nature of wind power and have a strategy for using alternative resources to meet your power needs

Conditions for grid-connected systems:

- You live in an area with average annual wind speeds of at least 10 miles (16 km) per hour (4.5 meters per second)
- Utility-supplied electricity is expensive in your area (10 to 15 cents per kilowatt hour)
- The utility's requirements for connecting your system to its grid are not prohibitively expensive
- Local building codes or covenants allow you to erect a wind turbine legally on your property
- You are comfortable with long-term investments

Although you can buy wind-power kits for your property, you really need to base your decision on information that is specific to your location. By using resources such as the Department of Energy's *Wind Energy Resource Atlas* and wind data from the National Climatic Data Center, you can determine if this option is a viable one for you. Keep in mind, though, that property may have obstructions or location issues. For example, if you are located in a gully or on the leeward (sheltered) side of a hill, you may not get the most benefit. Also, choosing a location for the turbine may preclude placement of future buildings in its path.

Wind power is readily available, and the supply is inexhaustible. It's completely free to use, it's low maintenance, and the energy generation is relatively low cost. However, there are drawbacks:

- Unreliability of constant energy flow
- "Idling" backup energy system, usually fossil fueled, needed
- Large physical space needed for power generation and battery storage
- Aesthetically, not pleasing in usually open, natural areas
- Birds and bats may be killed by blades

BIOMASS ENERGY

One of the most exciting forms of renewable energy is biomass, which can be used for lighting, cooking, space heating, water heating, refrigeration, and to fuel vehicles. Biomass is simply solar energy stored via photosynthesis; it comes from any animal- or plant-derived biological material. Plants most often used are miscanthus, switchgrass, hemp, corn, poplar, willow, sugarcane, and oil palm (palm oil). Biodegradable wastes are also used — even sewage sludge can be used in the production of biomass energy. Many advanced forms of biomass energy, however, are not yet fully commercialized for end users; for example:

- Pyrolysis: heating organic wastes in the absence of air to produce gas and char; both are combustible
- Hydrogasification: produces methane and ethane
- Hydrogeneration: converts biomass to oil using carbon monoxide and steam under high pressure and temperature
- Destructive distillation: produces methyl alcohol from high-cellulose organic wastes
- Acid hydrolysis: treatment of wood wastes to produce sugars, which can be distilled

There are, however, conversion methods with which you might already be familiar, most notably composting — you've noticed that your manure pile produces heat, right? Currently that's not "usable" energy, but in the future it may be used in a process called Flash Carbonization (see box on next page).

Other lower-tech forms of biomass energy production are anaerobic digestion (decaying biomass to produce methane gas and sludge, usable as fertilizer) and fermentation and distillation (both produce ethyl alcohol). Burning wood is also biomass energy production, something to consider when you are roasting your marshmallows over the camp-biomass-energy-production-fire.

As a renewable fuel biomass has frequently been referred to as "carbon neutral," but some greenhouse gas emissions may be produced when the fuel is used as a replacement for fossil

GHG Potency

Some greenhouse gases are more "potent" than others, notably methane. When you consider a gas potent, you might expect it to be strong smelling, and methane would certainly fit the bill. But the potency of a GHG does not have as much to do with the smell as it does with the gas's ability to infuse itself throughout the atmosphere. Smaller amounts of methane can spread out over a large area. As well, it's much more deadly than the other GHGs.

Think of it this way: Imagine you have 264 gallons (1,000 liters) of water. What happens if you pour 1 cup (237 mL) of milk into the water? It dissipates fairly quickly, but the water is still usable. What happens if you pour a cup of gasoline into that same tank? The water is completely contaminated. Our atmosphere can accept many kinds of gases, but some of them are more potent (that is, stronger or deadlier) than others.

FLASH CARBONIZATION

One of the most exciting new technologies for the agriculture industry is a process called Flash Carbonization, which was developed through research at the University of Hawaii at Manoa. The process superheats any biomass (even some synthetic materials, like shredded tires) and converts it to charcoal, bio-oil, and syngas, all of which can be used in environmentally friendly ways.

About 60 percent of what is produced is bio-oil, which can be used as fuel for large boilers and may one day replace coal as the fuel that provides electricity to the general public. The remaining mass that is produced is split equally between syngas and charcoal. Syngas is a combustible fuel that has half the energy density of natural gas and consists primarily of hydrogen, carbon monoxide, and some carbon dioxide. The Flash Carbonization system can be designed so it is powered by syngas.

And finally, we have charcoal. The uses for charcoal are many:

- Potting soil (for orchids and ornamentals)
- Cooking (barbecue) fuel
- Ultraclean coal (power production)
- Activated carbon (water treatment)
- Metal reductant used in the production of iron and steel (currently, coal is used)
- Soil amendment for field crops (which mimics the result of a naturally occurring prairie fire and rejuvenates soil)
- Biocarbon fuel cell (power production)
- H_2 fuel-cell electrodes
- Hg removal from coal-combustion flue gas
- Benzene removal from water

What makes this process most exciting from an environmental standpoint is charcoal's place as a sustainable fuel replacement for coal, which produces high levels of greenhouse gas emissions but also produces the lion's share of our electricity each year. According to the University of Hawaii, "Coal combustion adds about 220 lb (100 kg) of CO_2 to the atmosphere for every million BTU of energy that it delivers; whereas crude oil adds 170 lb (77 kg) per million BTU, gasoline adds 161 lb (73 kg) per million BTU, and natural gas adds 130 lb (50 kg) of CO_2 to the atmosphere per million BTU of delivered energy. On the other hand, the combustion of charcoal — sustainably produced from renewable biomass — adds no CO_2 to the atmosphere. The combustion of charcoal does not add to the atmospheric CO_2 burden because charcoal is produced from waste wood, crop residues, and other renewable biomass that would otherwise decompose (i.e., rot) in a landfill or in the ground and become CO_2. Thus the combustion of charcoal is a small part of nature's great carbon cycle upon which life depends."

So how does this affect you as an environmentally friendly horse owner? If your community had this technology, it could take every bit of dried manure and agricultural waste — even human and medical waste — and turn all of it into usable, green sources of energy. Keep an eye out for this expanding technology.

fuel and is not reclaimed in any way. When biomass is used for energy production, however, it is referred to as a "net reducer of greenhouse gases" because all by-products of its production (such as methane) are captured and used. Since methane is a much more potent greenhouse gas (see box, "GHG Potency," on page 29) than carbon dioxide, its capture in controlled combustion is very important.

GEOTHERMAL ENERGY

Geothermal energy is defined as any heat that is contained within the ground. The center of our planet is very warm, and heat radiates out of it. It's hard to believe in the middle of a Minnesota winter, I know, but in certain areas of the planet the heat comes closer to the surface. This energy, for example, is what heats natural hot springs. If you've ever been to Banff, Alberta, you might have visited the famed Banff Hot Springs, and Iceland and some European countries have become famous for their hot springs. Aachen, Germany, has the hottest springs of central Europe with water temperatures of 165°F (74°C).

Large-scale geothermal energy use (for electricity) will not become commonplace on farms in the near future for many reasons. Larger power plants are needed to convert the energy into a usable form, and the cost to produce and the space needed are also very high. But there are some appropriate applications, in the form of geothermal heat pumps. This system consists of pipes buried in the ground that draw heat into a heat exchanger and into a building via ductwork.

DEFINITION: Geothermal

The word literally means "earth heat." Geothermal energy is produced by heat stored deep underground.

Some parts of the planet are better suited than others for geothermal heating. In the United States, areas of Nevada, Oregon, Idaho, Arizona, and Utah hold high potential and have seen an increase in geothermal development. These areas have various geological conditions that allow water to circulate to the earth's surface and, most notably, the hot, dry rock that is accessible underground to tap into the earth's heat. This characteristic is present in less than 10 percent of the earth's surface land area. To harness this form of geothermal energy, the rocks are drilled and broken apart, then water is pumped down, causing the temperature of the water to rise as it comes in contact with the rocks. The water comes back up and powers turbines to provide electricity.

HYDROPOWER

Whether hydropower becomes a viable option for your farm depends, of course, on your location. The energy in water can be harnessed in many ways:

- Kinetic energy (the actual movement of the water)
- Temperature differences
- Salinity gradients

In fact, depending on the power company you use, you may already be receiving energy produced through hydropower — as are, for example, many residents of Ontario and Quebec. If you are considering microgenerating your own power, the relevant option for most farm owners will be kinetic energy, but only if you have access to a stream or river on your property.

If you have a source of running water, then you can purchase a commercial hydropower turbine. It works by diverting the water through the turbine and then releasing it back downstream. There will be a few factors you need to take into consideration, such as the volume and pressure of the water and whether they remain consistent enough throughout the year to offer a viable electricity option.

Just as with other microgeneration, the batteries will be the key to success. The larger your water source the better, as smaller rivers and streams may experience flow reduction during times of drought (summer) or may freeze completely in the winter. However, you can use these downtimes for equipment maintenance and restoration.

One of the best resources for building a small hydropower plant on your own property is the European Small Hydro Power Association. They have developed a guide (see Resources) to teach you some efficient ways of building and maintaining your system.

Depending on the size of the water source (smaller stream, larger river, waterfall, etc. . . .) your hydropower plant may have a large or small impact on the water, and this is something you'll have to consider. If you need to divert almost an entire stream to generate enough power, you need to think about the fish and other aquatic life that might be harmed. Some smaller plants can run in the water and allow fish and other creatures to pass right by the hydro generator intake, resulting in a smaller impact on the environment.

MAKING THE RIGHT CHOICE

In 2006 less than 20 percent of energy consumption came from renewable sources. Traditional biomass (wood burning) was at the top with almost 13 percent, followed by hydro at 3 percent. The more modern options — wind, solar, geothermal, and ocean energy — totaled less than 1 percent, even though the potential in these technologies is considerable. Wind-power use is growing by 30 percent each year (wind power currently generates 1 percent

of the electricity used in America), and the solar industry is just getting warmed up (pun intended). The largest solar thermal power station is the 354 MW SEGS power plant in the Mojave Desert. Geothermal power is also growing; the world's largest installation is The Geysers in California, with a rated capacity of 750 megawatts.

Producing ethanol fuel from sugarcane may become more popular in the future. This production figures prominently in Brazil's renewable-energy program (one of the largest in the world) and now provides 18 percent of the country's automotive fuel. The only downside is that sugar may become more expensive.

If you doubt whether rural and agricultural communities will embrace alternative energy sources, consider the nation of Kenya. It has the highest household solar-ownership rate in the world, with about 30,000 small (20 to 100 watts) solar-power systems sold per year.

We can do that. We can do better than we are doing now.

☼ ❄ CLIMATE VARIATIONS

There is incredible diversity between the regions in North America and even more diverse regions within regions. It may help to talk to people who have lived in the area for a long time to determine what long-term patterns may occur.

Subarctic

What a challenge you have! Most of us know that you do not live in igloos all year long, and in fact, the few urban centers have thriving equine communities. (As a child, my first-ever horse-back ride was here in this region upon a hairy old gelding owned by the neighbors in High Level, Alberta.) This region has a history of turning challenges into opportunities. Look for ways to use wind power generation year-round and solar panels in the months with more sunlight. You also have access to one of the most efficient water storage systems around: snow.

Humid Continental

The variable weather patterns that affect the humid continental region may make it difficult to choose just one form of alternative energy for your barn. You might consider using one or two complementary systems or a set of smaller systems, such as solar to run fencing and lighting in outbuildings and wind for barn energy.

Humid Oceanic

There is often a great deal of sunlight in your region, so solar power is a good choice.

Highlands

Wind power is a great idea for you if you live in the highlands region because your high elevation means wind speeds greater on average than the rest of the continent.

Semiarid

The options are pretty wide open: you generally have enough sun for solar, you can use wind power, and in certain regions you have access to great sources of geothermal heat.

Arid

A good deal of your personal energy will be spent trying to keep your barn cool and comfortable while minimizing costs and keeping everyone healthy. Maximizing water conservation is of key importance in any green barn, and this includes the trees you will likely be employing for shade. Make sure your water use is monitored regularly to maintain those green, leafy shade providers.

Something for Everyone

One of the great things about our continent is that we have an incredible variation in climate zones. In some areas we can surf in the morning and ski in the afternoon. This diversity means that there is never going to be one green solution that fits for every farm. Before you spend money to implement environmentally friendly power solutions, make sure that they are going to be the best fit for your farm and that the research you use takes your specific climate into consideration.

Also, think about the costs of implementation and the life cycle of whichever product you are choosing. How much will you use wind generation? Can you supplement with solar generation? Or, could you make do with a barn that utilizes natural lighting and does not require electricity at all?

UNDERSTANDING YOUR LAND

Planning and site selection

THE FIRST ASPECT of green horsekeeping we'll explore is the environment in which your horse lives. Planning and creating an environmentally friendly farm means incorporating the features and needs of the surrounding ecosystem to minimize impact. If he's an outside dweller, his paddock is his home, and it will be important to keep that land healthy as well, which is a topic we'll cover in chapter 9 (Field and Pasture Rotation).

In this chapter, we'll talk about the farm holistically: that is, we'll look at the farm as an entire system and how each part works in balance with the others to create a healthy or unhealthy environment. If one part is "sick," then the system as a whole is out of balance.

SITE SELECTION: THE BIG PICTURE

You might wonder why you have to put so much thought into the selection of sites for your barns, gardens, paddocks, and pastures.

First, in order to be as ecologically friendly as possible, you need to integrate your structures as seamlessly as possible into your environment. Choosing the wrong site may mean you have to make modifications to the land to shore up buildings, divert water, stabilize roads, and manage runoff, modifications that would be unnecessary if the right site had been chosen from the outset. You can use the land to your benefit and enhance its ecology rather than encroach upon it.

Second, choosing the right site will also mean you can make the most of any alternative energy sources and rely less on fossil fuels. For example, if you have located your hay stack on the opposite end of your property from your tractor storage and you have to drive the tractor across the property each day for hay, you are wasting fuel that could be saved if you had the stack or the shed located more conveniently. Or, as we learned in the previous chapter, if you have built your barn without taking into account the pathway of the sun, you won't be able to take advantage of passive solar energy as effectively as if you'd planned for it to begin with.

Mapping Your Land

Mapping your land is the way to start. Creating a bird's-eye view of your property with as much detail as possible gives you the best overview of how various sections of your property interrelate. As you'll soon see, the greenest buildings are integrated into their surroundings to cause the least possible impact. The buildings should enhance the ecology of your farm. When you build structures on your land, they should not disrupt or damage these natural relationships but rather should improve them.

The best resource for properly mapping your land is a topographical map, which can be obtained through your local planning office or county Extension office. In Canada contact your municipal district or county. Alternatively, you can have an aerial photo taken of your land, although it won't necessarily show all the detail you need, especially in areas where there are lots of trees that block

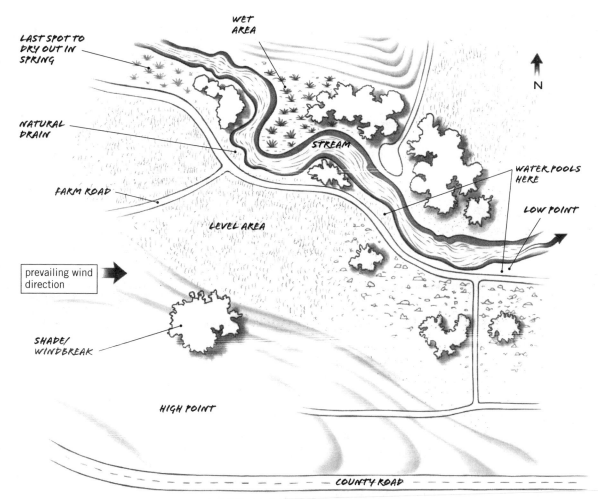

WET AREA

LAST SPOT TO DRY OUT IN SPRING

NATURAL DRAIN

FARM ROAD

STREAM

WATER POOLS HERE

LOW POINT

LEVEL AREA

N

prevailing wind direction

SHADE/ WINDBREAK

HIGH POINT

COUNTY ROAD

Mapping your property's natural features and resources is the first step to understanding it.

the elevations or dips in your land. In addition, counties and regional districts have a Registry of Deeds office with maps of all deeded properties in the area.

With map or photo in hand you can trace or copy the outline of your property onto graph paper. A real property report from your county will offer exact property line measurements.

Once you have the correct dimensions you may add the details of your property. The features of your land will determine how complex and accurate your map needs to be. If you have 10 acres of flat prairie soil in a nice little square parcel, you may only require a map that has the property lines drawn in proper proportion. If there are multiple landforms on your property, however — creeks, hills, rivers, roads, stands

of trees — then it will help to be as accurate as possible, referring to your topographical map or photo.

To complete your hand-drawn map, you should do a little footwork, mapping out the length of roads, fences, streams, and buildings and their distances from each other. It may not be necessary to know the exact degrees of elevation on your land, but you should note where high and low points are located. This will help you determine the direction of aboveground water flow during rainstorms and snowmelts, if applicable. If you have the time, walk around your property during heavy precipitation or melting. You will easily be able to see where the water is flowing. We've all heard the proverb "Do not build your house on shifting sand," but

MAPPING THE SUN AND WIND

You can find out everything about the sun — from its position and its track to the exact number of minutes it is up each day — at the Applications Department of the U.S. Naval Observatory (see Resources).

As for the wind, to help you determine which way it most commonly travels across your property, the National Climatic Data Center has a table that shows prevailing winds for many American cities. The table includes wind analyses from 1930 through 1996.

You can make some general observations from the data. You will notice that while prevailing winds in some cities come from as many as three different directions, in other places breezes blow just from one or two.

For example, in Montgomery, Alabama, the wind rarely blows in an easterly direction, staying mainly from northwest to south.

In Fairbanks, Alaska, winds are consistently from the north, except during June, July, and August, when the prevailing wind switches to southwest. Phoenix, Arizona, sees winds from the east for most of the year, except in September, October, November, and December, when it becomes westerly.

Some areas of North America have specific wind patterns associated with related weather events. For example, in the interior western portions of North America (mainly Alberta and Montana, east of the Rocky Mountains), a Chinook wind is characterized by a distinct arching cloud and drastic temperature increases. In January 1972 in Loma, Montana, a Chinook wind caused temperatures to rise from minus 54°F (minus 48°C) to 49°F (9°C) in just 24 hours.

have you heard the closely related "Do not build your barn on a puddle"?

Also include the direction of the sun, prevailing wind direction, and location of all water sources, including standing water and streams that run through your property. Note any natural and man-made drainage areas, as well as sections where water pools after a rainstorm and the location of the last spot to dry in the spring, as this might indicate either a high water table or a low spot on the property. This information will tell you where to build and what places to avoid. Knowing the slope of the land allows you to plan for natural drainage or water diversion if required.

Mapping solar patterns is an essential piece of research to do before building your barn. As you learned in the previous chapter, passive solar is one of the easiest ways to utilize natural energy. The critical element is to know where the sun is going to be and how you are going to take advantage of it.

Regardless of where in North America you live, knowing where the sun travels is key. Remember the phrase "The sun rises in the east and sets in the west, and all in the southern portion of the sky." It's probably one of the first phrases I ever remember my father teaching me (in an attempt to prevent my ever getting lost in the woods of northern Alberta). If you live in the South, you will want to know where the sun is going to be most of the day so you can build shade for the hottest parts of the year. If you live in the North, you will want to know the path of the sun so you can capture as much solar energy as possible for heating.

SOIL HEALTH

Rich, green, lush pasture starts underground in the soil. For the grass to grow, the soil must be able to hold the nutrients you put into it. With pastureland that is only being used to grow grass

---

DEFINITION: Dirt or Soil?

Dirt is actually displaced soil, meaning that it has moved from an area where there were plants growing in it, and it contained nutrients and had access to water. Dirt is what you find on your boots and what gets in your mouth on a windy day or what's in your arena footing. Soil, on the other hand is productive earth that has nutrients in it and is suited for growing. Recognizing the difference will help you maximize the fertility of your property. If it's on the road, your boots, mouth, or hair or blowing in the wind, it's dirt. If it's found in the grass and it's dark and rich and you could grow something in it — that is soil.

---

or possibly some legumes such as clover, your most important job will be proper management of the land, planning for the correct number of horses and the cycle of land rest periods. If your land has been used as a pasture or has experienced overgrazing or drought, it may be depleted of its nutrients, so determining the state of your soil is the first step to help you fix it.

The Natural Resources Conservation Service (NRCS) maintains all soil surveys and maps for the United States; you can rely on these reports to tell you the general type of soil on your property. Nevertheless, walking the land yourself and testing it will make you certain of the kind of soil you have to work with.

There are 14,000 different soil types in North America; there might be 150 different types of soil in just one county, and very often each small property will have two or three different types. Before you begin to put up buildings or fences or establish pastures, you should determine where the healthiest grazing lands are so you can make the best use of your pastures. Being able to use your pastures efficiently will reduce the amount of feed you need to bring in for your horses; chapter 9 will tell you exactly how to do that.

This type of foresight, integrating proper usage into your land planning, is an important part of being a green horsekeeper. Why build your barn on the best pasture, only to have to fertilize and encourage pasture growth in other areas of your property? Good pastureland often occurs naturally when a specific area has good hydration, healthy soil, and enough nutrients.

Mapping Your Soil

First, contact your Soil and Water Conservation District to get a copy of the Soil Survey Map. They will ask you for identifying information that is unique to your land, such as your specific parcel ID as noted on your land title or tax form, and will enter that information into a GIS (geographic information system) machine.

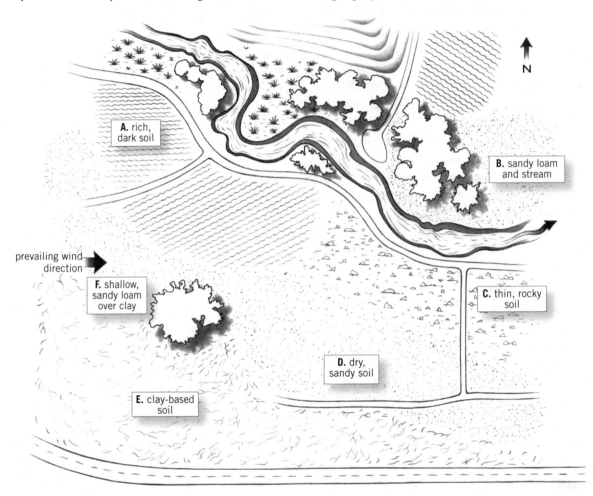

A – Ideal location for pastures because the soil is great for growing grass.

B – Look for sensitive riparian areas that may require protection and possible fencing. Also, check for erosion, especially if nearby soil is sandy.

C – This area is appropriate for a manure pile if it is flat and there is enough of an easement from the stream. Any fencing placed here will require deep, possibly metal, posts due to rocks.

D – An area with less potential for great growing soil is good for a sacrifice area.

E – This soil packs down easily and is a solid spot for building structures.

F – Paddocks might be located in an area with excellent drainage, where you can direct water away from high traffic areas.

They will then be able to tell you the soil types in your property. In Canada, contact your provincial department of agriculture or access the National Soil DataBase through the Canadian Soil Information System (CanSIS) division of Agriculture and Agri-Food Canada.

Knowing "what" your land consists of, however, is not enough. Now you can use the Soil Survey Map (which is often a large bound book, depending on your county) to determine the capabilities and potential of your soils. For example, softer, sandy soil will not be suitable for supporting large structures without making adjustments to the land. If the soil where you intend to place your barn is sandy and soft, you will need to bring in large equipment to remove the soil and rebuild the spot with harder, packed soil that could support a large building without shifting. On the other hand, if your paddocks and pastures are placed

in that same area, you would need to spend more time fertilizing it, possibly to no avail, as sandy soil does not hold nutrients well. A map of this type is as valuable as the résumé of a job applicant.

Testing Your Soil

A soil test will tell you what type of maintenance or nutrients your soil needs to reach its potential and how to maintain it at peak performance. Many Extension offices recommend comprehensive testing every three years to ensure that you are not damaging the land or depleting the nutrients through overgrazing. It is much better to adjust your fertilization every few years than to end up with declining pasture and forage quality year after year until your soil needs a major overhaul to bring it back to life. Some Extension agents will come test your soil free of charge.

THE THREE INGREDIENTS OF SOIL

There are many different terms for types of soil (sandy, heavy, light, good, poor, and so on), but there are three basic elements that all soil types contain: sand, silt, and clay. The proportions of these three elements will dictate how fertile the soil is and how it drains or retains water. Sand is a large particle that does not hold nutrients very well. Silt is smooth and powdery and slick when wet. Clay is also smooth but becomes sticky when wet, and while it can hold nutrients, it doesn't allow air or water to pass easily.

Loamy soil is the ideal type for agriculture and therefore would be a good soil on which to place your pastures. It will have a ratio of 40-40-20 percent concentration of sand, silt, and clay: enough sand and silt to allow water

and air to pass through and enough clay to hold nutrients and keep the soil from blowing away.

Here is a simple test to determine if you have loamy soil. Fill a glass jar one-third full of soil. (Note the location where you gathered the soil.) Fill the remainder of the jar with water, shake, and allow it to settle overnight.

In the morning, if you have sand-based soil, it will have settled to the bottom of the jar, and the water will be mostly clear. If the soil is loamy, organic matter will be floating in the water and the soil will be very dark. A clay-based soil will settle at the bottom, leaving a line of silt (or less-dense soil) between the clay and the remaining water.

First, you need to determine the pH of your soil because nutrient uptake is determined by the soil pH. If the soil is too acidic, then it doesn't matter how many different fertilization methods you use; the plants simply will not be able to access the nutrients. A pH of 6.2 to 6.5 is optimal for grass growth; around 7.0 is ideal for legumes, such as clovers.

You can buy a simple pH test kit from your county Extension or seed seller, and directions are provided with the kit. By mixing a selection of soil (see box, page 39) you will obtain an aggregate sample with the general characteristics of that area. When testing on your own, you will test each section of land individually. If there are already sections fenced off on your property, mark each sample accordingly (e.g., pasture 1, lower pasture, front pasture). You should also mark the samples on your property map for future reference.

SELECTING SOIL SAMPLES

Walk in a large W-shaped pattern throughout the area you have chosen, and stop every 20 steps to take a sample. For pH testing, you only need to go down about 6 inches (15.2 cm) or so. A simple tool can be constructed with a ½-inch (1.3 cm) pipe. Simply press the pipe down into the earth; when you remove it, there will be soil inside. Bang it on the inside of your sample bucket to dislodge the sample into the bucket. Avoid rocky, wet, or muddy areas when selecting samples.

For comprehensive soil testing, collect about one pint (about half a liter) of soil in the same manner as above to ship to a laboratory for testing. Your Extension office will have a list of labs closest to you. Comprehensive testing is more complicated than pH testing and not a do-it-yourself project. It will measure the pH, macronutrients (nitrogen, potassium, phosphorus), micronutrients (such as iron and boron), and the organic-matter content that affects

water and nutrient absorption and retention. The report given to you by the lab will include a recommendation for what to add to your land to bring it to its full potential, including how much lime (to raise pH), fertilizer, or seed is required.

Your Soil's History — and Possible Contamination

Soil testing can go much further than generic typing and fertility assessment. When you walk onto a piece of land, you might not know what has happened there or who has been on it. You can't tell from looking at it whether it's been used appropriately or abused in the past. And you may not be aware of contaminants making their way from the neighbor's land to yours.

It's been said that good fences make good neighbors, but fences won't block environmental pollutants, contaminants, or excess fertilizer. Know as much as you can about who your neighbors are and what they are doing with their land. It's not that you want to be a nosey busybody, but if they are doing something to cause damage to your land, you need to know about it so you can take steps to prevent it. Once contamination occurs, the road back to healthy land is difficult and (depending on the contaminant) may involve soil transplants and other expensive undertakings.

You should also be aware of what is located upstream from you: not just the neighbors but cities and towns and any factories or other possible points of contamination.

You can go to the Environmental Protection Agency Web site (www.epa.gov), choose your area of the country, and read the news on contaminants and spills or sign up to get those releases by e-mail. Currently, Canada does not have such a system through Environment Canada, though large environmental emergencies are often announced through television and radio broadcasts.

ASKING THE ESSENTIAL QUESTIONS

County Extension agents, real estate agents, and neighbors are great sources of information regarding your land's history. Here are some questions to get you started:

- **From whom was the property purchased, and what was it used for?** If you discover that it could possibly have been the site of any sort of manufacturing, farming, or dumping, then contaminants could remain.

- **What is the zoning history of this area?** Was it ever zoned differently in the past? Contact your county government and request a zone-history letter, which will tell you the zoning history of your land and whether a past use might have polluted your land.

- **What does your county Extension agent or your neighbor say about your land?** Ask around to discover the history of your land and whether there were issues in the past — some of which might be problems you hadn't considered. For instance, as scary as it is, in this day and age some country homes have been used for drug manufacturing because of their seclusion. The by-products of this type of "business" can be toxic. For example, methamphetamines produce five or six pounds (2.27 or 2.72 kg) of toxic waste for each pound of the drug.

A full-site analysis is quite expensive and involves extensive testing at different locations around your property. The best bet — and also the most cost-effective — is to focus on what types of chemicals are suspected. Find out all you can about the history of the property, including any activities that might have produced chemical by-products, such as tanning. If the property's main history is agricultural, however, focus your investigation on pesticides and petroleum products, as they will be the most common site contaminants.

What If You Suspect Fuel Contamination?

Q: *I purchased an old farm to be used as a horse-training facility. The farmer did not use a concrete pad around his fuel containers or special storage, so over the years a lot of fuel spilled on the ground. How can I determine the level of contamination, and what implications does this have for the rest of the property?*

A: The ground can be tested for fuel contamination, but this is not a do-it-yourself job. To test several feet deep, equipment will be needed. In addition, groundwater such as dugouts, sloughs, and streams will need to be tested. Soil testing to the depth of groundwater and down the slope (in the direction of groundwater flow) will be factors in determining whether any surface water would be impacted by soil contamination around the fuel-storage area. If the groundwater is contaminated, tests can follow the contaminated plume down the gradient to see where it goes. (See Resources for testing services.)

Q: *Can the area be cleaned up?*

A: There are many options for cleaning up petroleum contamination, and all are fairly cost-effective. The choice of a particular one will depend on the concentration of contaminant involved and the soil type.

Look for places on your property where there is an absence of plant growth in an area of fertility, such as a lush field that has a large bald spot for no discernable reason (for instance, a good reason might be that it's a popular gathering spot for livestock). One of the most common sites of contaminants on older farms is the burn barrel. Remember back when anything could be burned in the old burn barrel? Fifty years ago it was still okay to throw plastics and manufactured products into the fire. Now the Environmental Protection Agency has identified these sites as a "major" source of dioxin contamination.

Most land-grant universities offer free or low-cost soil-testing services to landowners, farmers, and gardeners that specifically test for toxins or contaminants. The process is very similar to the testing procedure outlined previously. When testing for contaminants, however, it's best to have the soil analyzed at a local or regional level because the lab will be familiar with the unique soil and possible contaminants in your area. Commercial labs can process your soil sample, and do-it-yourself kits for gathering the sample are available at most garden centers, county Extensions, or agricultural offices.

The National Sustainable Agriculture Information Service provides a list of alternative soil-testing laboratories in the United States. In Canada the National Land and Water Information Service provides many different resources for testing soils and water. (See Resources.)

Basic soil-analysis tests will evaluate nutrient levels and some chemical characteristics. Common nutrients tested for are nitrogen, potassium, phosphorus, calcium, and magnesium. The chemical characteristics assessed are generally the pH (acidity or alkalinity) and CEC (cation exchange capacity — a test of fertility) of the soil. The ideal pH for most crops and grasses is 6.5 to 7.0, although there can be variances due to the levels of various nutrients.

Soil tends to acidify over time, especially if there has been overuse of fertilizers or if land has been continuously planted with crops over an extended period of time. This will affect the rate of growth for your pastures. If you believe you are managing your pastures correctly (you'll learn more about this in chapter 9) but you are still having issues with consistent, healthy grass growth, soil testing should be your next step.

MAPPING WATER

Water is the lifeblood of your property. In chapter 8 we will discuss its importance and how to protect it in great detail, but our goal in this chapter is to help you understand water patterns. As you build your barn, fences, and other structures, this understanding will guide you in protecting the water system and not disrupting its flow or quality.

While mapping your property, pay careful attention to the direction of any aboveground water flow and the grade on either side of it. Simple arrows indicating the direction of flow as well as the grade will help you predict and eliminate possible sources of contamination from paddock, road, and building runoff.

It will also be useful to know in which direction underground water will be flowing. As you may know, the Great Divide runs from Alaska down to Mexico. Water on the east side flows toward the Gulf of Mexico, while the western water flows to the Pacific. There are also three other divides: the Northern (Laurentian), the Eastern, and the St. Lawrence Seaway.

These divides direct water in various directions toward water basins such as the Great Basin, Hudson Bay, the Labrador Sea, the Gulf of St. Lawrence, and the Mississippi River and are not just important with respect to water flow but also figure in political and legal issues. For example, water in North Dakota drains into Manitoba, Canada; therefore residents in Manitoba are often concerned about water regulations and possible contaminants from another country.

By understanding in which direction your underground water flows, you will know whether to be concerned about any production and agriculture that is located upstream from your property. Contaminants put into the ground can make their way to your farm and are very difficult to extricate, so it's advisable to be vigilant and know all the laws and

NORTH AMERICAN DIVIDES

Arctic Ocean

Hudson Bay

Gulf of St. Lawrence

Pacific Ocean

Atlantic Ocean

Gulf of Mexico

——— Great Divide

••••• Eastern

••••• Northern

–•–•– St. Lawrence Seaway

The North American Divides (the Great, the Northern or Laurentian, the Eastern, and the St. Lawrence Seaway) determine the flow of water through our continent.

regulations for your region. Also, you should be aware that contaminants from your own farm can make their way downstream to your neighbors' lands.

DETERMINING LAND USAGE

Now that you have identified the various features of your property and their locations, your next step is to determine which areas of land to reserve for which specific purpose. Note on your map any areas that are already spoken for,

such as where the barn or home is or will be situated. Use the natural curves and land formations to create natural boundaries on your map. Barriers such as roads, streams, and valleys can become map lines, as they will provide natural barriers and help determine how the land should or could be used. You won't likely use a sloped field with a stream running through it as an outdoor arena or paddock.

You can simply draw lines on your map and label them in whichever manner is logical; for example, you might have sections simply

Creating a Legend

Marking off your map will be easiest if you determine the symbols you will use ahead of time and stick to them. Create a key or "legend" for easy reference. For example:

- Solar patterns drawn in yellow, as arrows or dashed lines
- Wind patterns drawn in blue, as arrows or dashed lines
- Soft or sandy soil indicated with hatch or hash marks (#)
- Roads drawn in black
- Flowing water drawn with a combination of "squiggly lines" and arrows to indicate direction (~~~>)
- Underground water direction indicated in large blue carets to indicate direction (< or >)

numbered or lettered. Or you might determine names that suit the land formation, such as "front field," "south riverbank," "east slope," or "ravine."

Create a legend (see box at left) to ensure that you are consistent with your map markings. You may want to create several color copies so you can sketch out several possible uses. Don't forget to note on the map where utility poles and your utility hookups are located. If they have not been built yet, indicate the direction from which they will be coming.

For the remaining sections of land, write down every conceivable use that you want and may want in the future on a separate piece of paper. The chart below lists some ideas to get you started.

Some land uses won't fit snugly on this list, so you may need to be specific. For example, you can't fit "breeding/raising horses" onto a map,

🍃 LAND USE WISH LIST 🍃

ACTIVITY OR FACILITY	MUST HAVE	WOULD BE NICE	MAYBE SOMEDAY
Boarding horses (how many?)	☐	☐	☐
Competitions	☐	☐	☐
Indoor arena	☐	☐	☐
Outdoor arena	☐	☐	☐
Round pen	☐	☐	☐
Trails	☐	☐	☐
Breeding/raising horses	☐	☐	☐
Growing crops	☐	☐	☐
Gardening (what will you grow?)	☐	☐	☐
Storing machinery	☐	☐	☐
Composting manure	☐	☐	☐
Fishing	☐	☐	☐
Raising other animals (kinds?)	☐	☐	☐
_____	☐	☐	☐
_____	☐	☐	☐

but you can include: foaling shed, weaning pen, breeding shed, mare and foal turnout.

Once you have completed your list, it's time to determine how you can accommodate the most important things on it. Begin to section off areas on your land map, starting with the "must haves" and include existing structures that will remain. It will be handy to have several color copies and a pencil to shade the areas. Though we're starting from scratch here in building your barn and arena, you might have some structures in place already, such as hay sheds and paddocks. You might choose to use these structures for the same purpose or move them to another, more convenient location. Then move on to the "would be nice" category and map out areas for each item you'd like, such as a hay shed to replace your "stack and tarp" storage method.

The last column of your table will be for items you'd like to add in the more distant future. Perhaps someday you'd like to add a

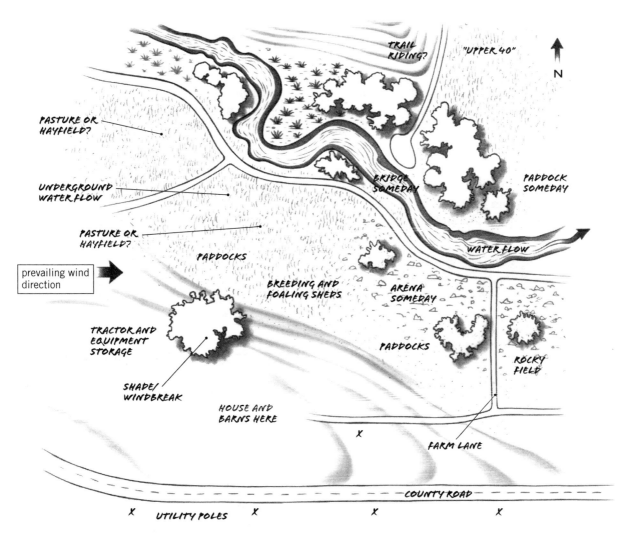

Whether you have structures yet or not, at this stage of planning you can let your imagination soar. Consider the ways you can use different sections of your land, keeping in mind the all-important realities of wind direction, slope, water flow, and access.

stocked fishing pond or riding trails to an unused portion of your property. Earmarking this land for future use will help you consider the effect other construction or use may have on it; you wouldn't want to put a fuel-storage tank in an area you have earmarked for a future vegetable garden. You can choose to note these on the map as future plans, or you can wait to see if you have the space for them.

Now you can start to choose the best locations for your barn, paddocks, and pastures. You may have some pastures already, but be open to using those areas for other structures if they are the most appropriate locations.

Drainage

Before you can decide what is going to go where, however, you need to ask yourself where the water is and where it's going. It is vitally important to find out as much about the property's drainage patterns as you can before you build. Each property will have a natural drainage pattern, and of course, the best choice for your property is to build around the natural drainage without disrupting the flow. If it is not possible to do this, however — for example, if your parcel of land can be accessed by only one road and the land naturally drains in a line parallel with that road — you may need to disrupt that flow. (Or buy hip waders.)

On the other hand, if you know that every spring, for instance, rainwater will travel from a hilly spot on your property, cross the northeastern corner, and flow across a proposed road, then you can avoid problems by implementing a workaround solution. One choice might be putting a culvert under the proposed road to allow the water to continue to flow naturally while still giving you a usable, dry road in the spring.

Drainage on your property is something that may need to be assessed over time. Certainly it can be hard to tell at first where things might

Need Help?

Environmental landscape architects can offer a wealth of information. If you don't feel confident doing the work yourself, hiring one as a consultant or to do the entire job for you is an option. You can find environmental landscape architects through associations and companies listed in appendix C of this book.

drain until that first big snowmelt, when you discover that several of your paddocks become unusable because they are flooded.

But take note of the natural ditches on the property. Look carefully at features at the bottom of each hill and know that when the rain comes, that area will be the muddiest. Will horses have to live there? Are there waterers, paddocks, doorways, or roads at the bottom of the hill or even halfway up one?

Many people have to fight water during rainstorms or spring melts. Those who haven't planned their property correctly can be seen with rubber boots and shovels out in the mud, madly trying to divert streams to flow to less damaging areas.

DECIDING WHAT GOES WHERE

The next step in your land-mapping process is to begin placing the fences and structures on your property. It will help to have several copies of your map so you can draw and redraw the proposed fence lines or move structures around. You might consider using strings and cut-out cardboard or sticky tag to represent fences and buildings. If you have an idea of your barn's dimensions, a little cardboard cut-out of it is easier than drawing it repeatedly. Have fun! It's a heck of a lot easier to move that barn on paper.

Locating Your Barn

Locate your barn at one of the highest points on your property, to ensure that water does not flow toward it or settle around it. Such a water pattern will cause issues not only with mud (which turns into dust in the summer) but also with the building's foundation. Over time water can be a powerful force, moving dirt and thereby possibly causing small fractures in a barn's foundation.

As John Wagner states in his book *Barns, Sheds and Outbuildings: Plan, Design, Build,* you should position your barn so that "the positive pressure from wind will drive cold air into the barn, and negative pressure on the far side will suck warm air out of it." While you are considering the orientation of your barn, don't forget to have the wall with the most glazing facing the south so that solar heat can add warmth if required.

COLD-CLIMATE CONSIDERATIONS

In a cold climate, your barn should be oriented along the east-west axis to maximize solar exposure to winter sun. This leaves two long walls, one facing south and one facing north. The northern wall should have more mass, thicker walls, or even earth moved up against it. This will help stabilize temperatures in both winter and summer, as the earth maintains a consistent temperature between 50 and 60°F (10 and 15.5°C). During the winter more mass on the north side will help keep the interior warmer, and in the summer the interior will stay cooler. If you are unable to "berm" the wall using the side of a hill (actually building the barn into the side of the hill), then you can use straw bales to do the job or, as mentioned above, move some dirt against the north wall.

DEFINITION: **Berm**

A mound of earth or sand.

If you are considering solar panels for micro-generation, choose the southern side of the roof for installation. The majority of windows should also be placed along this side if you plan to use passive solar for heating, as outlined in chapter 2. Dense evergreen trees can provide a good wind break along the north side or they can block the prevailing wind. Along the south side of your barn you can plant deciduous trees; in the winter their bare branches will not obstruct the sun, and in the summer the leaves will provide shade.

Having smaller spaces inside your barn will promote heat retention, especially when used in conjunction with good insulation (more on that later in this chapter). Many smaller barns in northern regions do not require any heating at all, because the warmth created by the horses inside at night is enough to keep them comfortable. Larger barns, however, need to maximize the heat of the sun to take the edge off the cold temperatures in the middle of winter.

WARM-CLIMATE CONSIDERATIONS

As you move farther south, the considerations of temperature change. In the deep South, heat can be a great concern — many more horses each year die of heat-related illnesses such as heat stroke and dehydration than of cold-related ailments like hypothermia.

In hot climates you should also build your barn along the east-west axis, but it's the southern wall that must be bermed with hay bales or earth walls to help your barn maintain a stable temperature. No insulation is required on this wall, though some sort of vapor barrier should be placed there to prevent condensation.

Most of your windows should be located on the northern side of your barn and fitted with shades or roof overhangs to allow light but not solar heat to enter. Open floor plans promote cooling, so if you have several stalls inside your barn, consider having slatted rather than solid stall walls. You can also create a "thermal chimney," a tall, vented space that directs heat outside the barn. Heat will rise up and out if it has a place to go.

SOIL

The soil where the barn sits should be clay based, as this allows for the least amount of water to penetrate. Clay also packs down firmly, providing a stable base, and does not erode as easily during the wet season. If your land has a high percentage of sand, you may need to remove some soil and fill in the barn site with packed clay.

WIND

Wind can cause havoc in so many ways: doors that wear out from constant strain, driveways and roads that are drifted in by snow during the winter, feed or bedding that is lost by being scattered around the property flake by flake. If your barn has a center aisle, plan it to be perpendicular to the prevailing wind direction. If the aisle runs parallel with the wind, whenever you leave your barn doors open for a little ventilation you'll get a wind that whips down the center of the aisle but doesn't pull air from the stall areas, leaving that air stale. If the wind blows across the barn, it will pull air from inside the barn and create natural ventilation.

If possible, use naturally occurring clearings and windbreaks to your benefit when building your barn. Structures located near your barn, such as hay or bedding sheds and paddocks, should also be placed in areas with natural or created windbreaks. If you are waiting for trees to grow, you can create a simple windbreak with a row of round bales.

Selecting Trees for Shade and/or Windbreak

Where you live in North America will determine what type of tree will thrive and grow quickly to provide long-term shade and windbreak as well as a haven for local animals and birds. Your state or provincial forestry service will be the best resource to use for selecting trees to plant and which ones will thrive in your particular environment.

There are certain considerations you must take into account.

TREE FUNCTION

Where is the site located, and what purpose is the tree going to serve? Shade trees block solar energy and are very helpful in reducing energy demands in the summer months. Border trees provide privacy and reduce wind while also blocking out unwanted noise. Street trees (green ash, linden, honey locust, and pin oak, for example) are called that as they are often planted on city streets — they don't create a visual barrier (to traffic or street signs) but they do reduce air pollution, noise levels, flooding, and soil erosion, and they filter dust. They may be appropriate along roadways that pass through your property.

SITE SPECIFICS

Are there any limitations to the site you have selected? You may want a large, leafy tree for shade and beauty, but your soil and the level of precipitation may not support it, as is the case in some of the more arid regions. If you plant a tree with high water needs in an area that often experiences drought, you might spend a lot of time watering it.

Will your tree be able to grow to its potential? The growth rate will be determined by soil fertility, rainfall, temperature, and age and species of the tree but might also be affected by overhead utility lines or other structures that could interfere with future tree growth. If the tree is oversized for the space it is in, you will need to prune it heavily or top it (cut the top off) regularly. The branches may grow lower to the ground and restrict passage underneath the tree or may be broken easily during tumultuous storms.

Other space considerations take place underground. Will the tree have room for root, branch, and trunk growth? Check with your county Extension agent about appropriate species for your region and their particular needs.

AESTHETICS

Will the tree fit into your environment? Having a tree that is pleasing to all five senses is a bonus. If there are spring blossoms or bright fall foliage or if the tree branches allow for safe passage beneath them, the tree will have a positive impact on all life on your property.

SAFETY

Some trees are considered toxic to horses and other livestock:

- **Walnut.** Every part of the tree is toxic, from flowers to pollen to root: exposure may lead to laminitis. The roots excrete juglone, a chemical that acts as a natural herbicide, killing off many nearby species of plants.
- **Native red maple.** Eating wilted red maple leaves causes hemolytic anemia, though the exact toxic ingredient is not known.
- **Cherry species.** Leaves contain prussic acid, a form of cyanide.
- **Black locust.** Seeds, leaves, bark, and twigs are toxic to livestock and humans.
- **Yew.** Leaves, seeds, and twigs contain a toxin known as taxine, which can cause collapse and sudden death; only half a pound (0.3 kg) for a 1,000-pound (453 kg) horse can be deadly.

Locating Your Arena

Indoor arenas require the same forethought that barns do for their location, lighting, and wind cover. Lighting is particularly important for safety. It is a huge benefit to have light sources that do not throw bands of scary light

---

Toxic Plants

If you are unsure about poisonous plants on your property, you have several resources available:

- University Extension services (either a veterinary or a botany department)
- Bureau of Land Management
- Local large-animal veterinary clinic
- Local office of the ASPCA

Universities are often very useful in helping identify toxic plants, as students in equine science or botany studies may need to do field examinations for credit. If you are concerned about the toxicity of plants on your property, call and speak to a department coordinator to see if there are students interested in coming out to study your field.

---

on the footing and make your horse believe a large snake lies in wait for him every afternoon at 4:30 when the sun hits just right.

Outdoor arenas are a bit trickier to place than barns. Again, they should be at a high point on your property. Ideally, the only water you want on your outdoor arena is what falls from the sky and what you place there on purpose. Many poorly planned outdoor riding rings are beset with a period of uselessness each year when they become flooded or soggy.

NO TREES, PLEASE

Outdoor arenas (and roads) should be situated away from large trees to protect the roots. Each tree has an intricate root system that branches out just like the branches from the crown. Some, such as pine trees, have more solid, vertical roots that anchor the tree deep within the ground. Others, oak and maple, for example, spread their roots wide.

At the end of the roots are fine hairs that help to draw water and minerals out of the ground. If these roots extend under an arena, over the years compression will damage those delicate root hairs, making it more difficult for the tree to take up nutrients from the soil. If you are located in an area with high rainfall, the roots of your trees are more likely to be near the surface. If the climate is drier, the roots will grow deeper, seeking the water.

SOIL AND DRAINAGE

A well-designed outdoor arena has drainage around it to direct water. In addition, it needs protection in place to keep the good footing where it should be so you don't have to replace all the sand after a big rainstorm. A popular option is to have a compact clay base with moderate to heavy sand on top. The clay provides a flat riding surface and does not let water permeate, allowing it to drain off. You can grade an arena base with a slight incline to the outside, allowing for drainage. The sand on top allows for a softer riding surface, and because the sand particles are large, they do not easily blow away when drying, nor will your arena turn to mud.

WIND EFFECTS

Naturally, outdoor arenas are affected by wind to a greater degree than indoor arenas are. While avoiding putting trees too close to the arena, you should try to find an area with natural windbreaks or create some by planting trees or creating a windrow with round hay bales. On the plus side, the wind plays a big part in helping to dry footing after a rain.

Indoor arenas require similar airflow considerations to those of barns. If you live in a climate that does not require heating and insulation, you should consider building solid walls about three-quarters of the way up to allow for airflow into the structure but not directly onto the horse and rider.

SOLAR INFLUENCES

An indoor arena can easily become unusable when the temperatures soar and the structure becomes much like a greenhouse. I have ridden in arenas where my horse broke into a sweat after simply walking around and my clothing stuck like wet tissue. There must be a balance between having enough light and blocking solar energy from cooking the inhabitants. Techniques may include:

- Using overhangs to block the sun in the hottest times of the day (generally from 11:00 A.M. to 3:00 P.M.), while keeping windows open
- Promoting air movement throughout the arena by opening east and west doors
- Using fans
- Utilizing building materials that will not absorb heat (for instance, light-colored walls and roof)

LIGHTING

Lighting is a serious concern in Canada and the northern parts of the United States, where riders often ride indoors during times of the year when it may get dark at 5:00 P.M. or earlier. An excellent option for these individuals is a steel-framed fabric-membrane building like a Cover-All Building System. The fabric membrane is wind resistant and allows a lot of light in. During the day you do not need lighting to ride, but you can add lighting as required, just as with any other building.

WATER

Dust control is very important in an arena, whether it's indoors or out. Therefore you must have access to water to periodically dampen down the arena footing. Be sure to provide a water source in your construction plans, especially for an indoor arena. You may want to consider using gravity-based collection methods that utilize your roof to collect rainwater to distribute over the arena footing. Information on creating a rainwater-collection system will be discussed in chapter 5.

The only water you want on your outdoor arena is water that falls from the sky or that you put there yourself. Consequently, you may need to divert building or land runoff away from the arena. This can be accomplished easily through grading the base (a raised midline allowing for runoff to the long sides) and digging a narrow ditch around the arena filled with gravel to divert any runoff coming into or out of the arena.

Locating Paddocks and Pastures

For the sake of clarity, let's differentiate between paddocks and pastures. A paddock is an enclosure that houses one or two horses but does not have enough grass to sustain the horses' feed requirements. A pasture, on the other hand, is large enough that it may have enough grass and land to feed horses throughout the summer months. And I say "may" because any pasture, if improperly managed, will eventually become overtaxed and be unable to support the nutrition needs of the horses. Of course, you can feed your horses hay or other roughage when they are on a pasture, but it will still be significantly less than you'd need to feed a horse in a paddock.

Finally, daily turnouts are often smaller than paddocks as they are used only for — surprise — daily turnout. They are located close to the barn for convenient turnout, and some may actually be attached to the barn, with turnout directly from the stalls.

You have many considerations to account for when you choose where your paddocks, pastures, daily turnouts, and sacrifice areas will go. Pastures need to be located where there is potential for growing good, healthy grass, so the soil should be fertile. But there should also be good drainage so that water does not collect in hard-

working areas; for instance, around gates, corridors, and waterer and feed locations. It is even better if your pasture has natural windbreaks, such as a stand or two of trees or brushes.

Gates should be located on a flat plane so that you and your horse can have sure footing when entering or exiting the field. See pages 53 to 55 for more on fencing.

WATERERS

This is also the time to decide where your waterers will be located. If you choose to have automatic waterers, you will need to plan where to lay water pipe. The waterers do not need to be located close to the gate but should be along the closest fence line and accessible by vehicle in case you need to bring equipment in for installation or repair. Do not locate a waterer (or any feed bins) too close to the gate, as this will cause horses to congregate in the area, which makes it more difficult to come and go through that

Paddock Don'ts

There is no one right way to build paddocks, but there will be configurations that you should avoid, such as:

- Long daily walks for turnout, especially with multiple horses
- Dragging water hoses
- Walking horses through narrow passageways (between paddocks or buildings)
- Feeding more than one horse in a small area
- Having narrow entrances to paddocks and pastures, especially where multiple horses are housed and the likelihood of conflicts is increased
- Having narrow alleyways that prohibit movement of equipment

entrance. With a waterer you run the risk of overflows creating a mud pit in your gate area.

DRAINAGE

Paddocks where horses will live during the day and night must also have good drainage, especially if you do not yet have a barn or if you have more horses than stalls during muddy periods. It's the poor, afflicted horse (usually one or two in a group) who gets shuttled to the "muddy paddock" during the spring and ends up with feet that are too soft or legs that get "scratches" (see box on page 53) from being constantly wet.

Any place a horse is going to stand for periods of time should not be prone to developing mud holes. As we'll see in coming chapters, mud in the spring is dust in the summer. As with pastures, your paddocks will benefit from natural windbreaks.

We will discuss maintenance of paddocks and pastures in greater detail in chapter 9.

TURNOUT AND SACRIFICE AREAS

The considerations for pastures and paddocks also apply to daily turnout areas, which are usually located much closer to the barn for ease

A muddy paddock is a recipe for future troubles, from hoof issues to respiratory ailments. Choosing the right location — a dry, well-drained site with wind and sun protection — will go a long way to head off problems before they begin. Weekly manure cleanup, coupled with appropriate bedding, will allow your paddocks to stay clean, dry, and healthy year-round.

of use. Unfortunately, this means that often they end up on the downward side of the highest point in the property: your barn site. If possible, keep your turnout pens on the high plateau as well, or consider some drainage options such as grading, building ditches to divert water, and building gutters and rain-collection systems to catch the water runoff from buildings before it reaches the turnouts. (Rainwater-collection systems will be discussed further in chapter 5.)

A sacrifice area is a place that is fenced off and used to house horses during wet periods. Consequently, this area does not have good grass growth because it is trod upon during muddy periods and the grass roots are damaged. The area is "sacrificed" for the good of the rest of the pastures, paddocks, and fields. We will discuss this topic in greater detail in chapter 9. During this planning stage, you can identify possible sacrifice areas. As you will read later, you can take a less-than-ideal location and add permanent types of bedding to prevent excessive mud buildup.

Fencing 101

Designing your pasture fencing and gates will go hand in hand with deciding where they are located. Ask yourself:

- Will I be walking my horses to and from my pastures each day?
- Will I be installing runs from the barn to the pasture so horses can have access?
- Will the pastures be used mainly for long-term, year-round housing or daily?

PLANNING YOUR FENCING

As previously mentioned, your property might have different types of soil in different areas. This can cause complications with fence construction because not all ground is good for all types of fencing. You would find it a futile exercise to try to pound wooden posts into a rocky field and might have to consider steel posts instead. Conversely, another section may have softer, sandier ground and require longer posts for stability. You can predict some soil types by looking at land formations. For example, you will find that the bedrock on your property is closer to the surface on top of hills and higher elevations.

When planning your fencing needs, it is essential to walk the ground. Know the soil and plan the type of fencing accordingly. Some areas may require 10-foot (3.1 m) posts at shorter intervals because of the softness of the ground, whereas harder land will accept 6-foot (1.8 m) posts with longer sections.

FENCING FOR EFFICIENCY

Now is the time to start planning where to build your fences so that you can operate at the highest level of efficiency. A great deal of time and energy can be wasted if you are constantly backtracking across your property during feeding, turnout, or maintenance. If you build your fences in a manner that allows you quick and efficient use, you can save yourself time and

---

DEFINITION: Bedrock

Bedrock is the topmost layer of solid rock beneath land formations. In fact, the movement of the bedrock has created those land formations.

---

maintenance. By building a fencing system that takes horse behavior into account, you will have less long-term damage to your property.

Here are some suggestions:

- Attach turnout paddocks to the barn to allow direct access from individual stalls. This means that your horse's stall actually performs double duty as a stall at night and a shelter from the sun during the day.

- Feed horses along the long side of a rectangular enclosure to give more room between each individual horse and to minimize fighting.

- Place gates in the middle of a short side of an enclosure rather than at the corner to minimize the friendly (and sometimes not-so-friendly) contact between neighboring horses during entrance or exit.

FENCING ACADEMY

Here are some popular fences, all of which have pros and cons (see chart opposite).

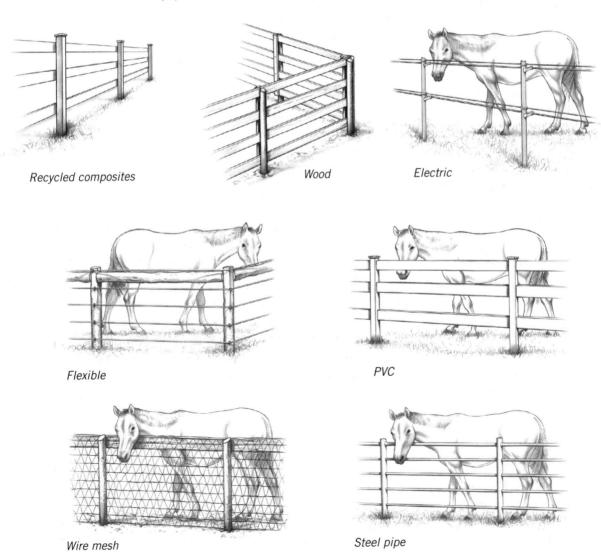

Recycled composites

Wood

Electric

Flexible

PVC

Wire mesh

Steel pipe

- Group paddocks (turnout or permanent) in clusters near the barn exits. For example, if your barn is a long rectangle, you may have two doors on either end of the alleyway. With two sets of paddocks you can locate water at both ends as well as move efficiently during morning turnout.
- Use wide alleyways between paddocks or cluster your paddocks so that there is a large area in the center to be used for maneuvering tractors (for paddock cleanup or single round-bale delivery). An octagon shape can be effective, with each gate fence opening into the center. Alleyways should be wide enough to drive a tractor between fences safely with several feet to spare. Anyone who has ever driven a tractor with a round bale down a narrow paddock aisle and tried to get the horses to stop eating from the "moving buffet" will appreciate the extra few feet.
- For daily turnout horses, always consider how fast you could get them inside to shelter if a storm came up quickly.
- Locate paddock and pasture shelters close enough to the gate that you do not have to walk all the way across the pen to retrieve a reluctant horse.

GREEN FENCING AT A GLANCE

FENCE TYPE	PROS	CONS
Wood	Environmentally friendly, especially when using wood from sources that replenish and replant forests; no damage to environment if thrown away or burned.	Higher maintenance, horses may chew
Electric	Psychological and physical barrier; good for temporary fencing. Electric-braid fencing is your best environmentally friendly choice because it is made of braided polyester (friendlier than the traditional PVC coating) and can be run off solar power quite easily (many kits are sold with solar boxes).	Does not make good permanent fencing when used with weak poles; may be used in conjunction with heavier posts
Flexible fence with high-tensile wire	Easy to maintain; has a lot of "give" if a horse leans against it; can be repaired easily by restretching.	Coated wire is encased in a plastic that is difficult to recycle, while uncoated wire may be dangerous to your horse's legs
Wire mesh	Galvanized wire is rust resistant and low maintenance; mesh openings are 2 inches by 4 inches (5 cm by 10.2 cm), so are too small for a horse's hooves.	Once bent out of shape, it can be difficult to restretch
Polyvinyl chloride (PVC) or vinyl-coated wood	Durable and cheap; easy to maintain.	Dioxin (derivatives increase likelihood of cancer) is a by-product of PVC manufacturing
Steel pipe	Very strong; low maintenance; some pipe corral fencing can be moved in panels, allowing for temporary and permanent fencing; a good investment for long-term use.	Steel mills use a lot of energy and fossil fuels to produce the pipe
Recycled composites — fly ash	Uses recycled material; very environmentally friendly, low-maintenance, strong, resistant to water absorption; can be treated like wood (cut, nailed).	Slightly more expensive than wood; horses may chew

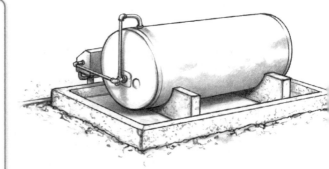

A fuel tank should be set atop a proper pad with cement barrier.

LifeTime Lumber

This ecofriendly wood alternative that uses 60 percent fly ash is produced by LifeTime Composites LLC in Carlsbad, California. The production plant boasts zero harmful emissions, and the energy-efficient manufacturing process consumes a minimal amount of water and electricity compared to other products. Fly ash is a nonhazardous mineral recovered from the combustion of coal in electric utility plants. The rest of this product is polyurethane, which provides a unique combination of weight, strength, hardness, flexibility, and resistance to water absorption.

Know when to choose safety over efficiency. If you employ a long fence corridor to let your horses meander from the barn out to the pasture on their own, you may find yourself with damaged gates or horses as they attempt to pass each other or engage in play (or fighting) en route.

Locating Storage Sheds

Each property has its own needs for storage and other buildings that are erected based on convenience, safety, and logic.

Storage is a concern for both efficiency and effectiveness: the location or building must be useful and do its job, and it has to be large enough to hold an appropriate amount of material for the proper length of time. Let me explain further: Having the world's largest hay shed is great but not if you fill it up and can't use up the hay before it starts to go moldy. If you do not build enough shelter, on the other hand, you may be forced to accept a new load of hay each week, increasing your reliance on the fossil fuels that the delivery truck burns. Managing your feed supply will be discussed further in chapter 10.

Here are some storage considerations for you to ponder.

FUEL STORAGE

Fuel should always be stored as far away from a water source as possible and always with an appropriate barrier (such as a cement pad) to prevent fuel spills from contaminating ground or surface water.

For the sake of efficiency and safety, it is helpful to locate fuel near equipment storage but away from barns or houses. We'll discuss contamination prevention in chapter 11.

BEDDING

Bedding (whether straw, pellets, or shavings) needs to be kept sheltered from the elements to prevent it from being blown away or ruined by rain or snow. Wet bedding cannot soak up urine in your stalls when it is already saturated.

HAY STORAGE

Hay is a precious commodity, which is why many horse owners choose to build hay sheds to protect the feed from rain, wind, and other elements. If you do not want to build a shed, you can simply tarp the hay and pull back the tarp as required.

PLACING BUILDINGS FOR EFFICIENCY AND CONVENIENCE

An often-overlooked area of design is the energy you need to use to move feed, bedding, and other farm necessities around your property. Just as you will plan your fences so you are moving horses around in an efficient manner, you should also examine the essential items you use each day and ensure that they are placed in efficient, accessible locations. Feed and bedding are the most common items that need to be shipped in regularly. The size of storage for these items will depend on several factors:

• How many horses you have
• Your climate (is it conducive to storing feed, or is it too humid, which will lead to mold issues)
• Environmental concerns, such as rodent or insect problems
• The availability of feed (for instance, your location may dictate a set delivery date from suppliers)

The most important aspect of hay storage is flooring. Hay that sits directly on the ground will eventually become moldy. Even before you can see the mold, it will be there, waiting to be eaten by your horse, and then it can wreak all sorts of havoc on his digestive system (colic being the most serious). The best option is a floor made of wood pallets, whose slats allow air to circulate underneath the stack of hay. Pallets are often used in transportation and made with lower-quality wood 2×4s that are not up to construction grade. They can be easily repaired if need be with hammer and nails and are easily recyclable.

Hay can be stored long term (for at least one year) without concern for spoilage if you care for it correctly. Other feeds such as grain and cubes will not last as long. Because these feeds need to be kept in a sealed container to prevent insects and rodents from entering, they can spoil more easily due to moisture buildup.

Having smaller containers for feed or smaller hay stacks ensures that you will use the feed efficiently and before it can spoil; however, you must also take into account the "cost" of having feed trucked in more often. Bedding can also be stored longer as there is no immediate danger of spoilage. Try to keep your deliveries to a maximum of one per month to save the environment from excessive CO_2 emissions.

The feed-delivery system you use on your farm will vary depending on the size of your property. Small-acreage owners can throw a hay bale into a wheelbarrow, walk around to their outdoor pens, and feed their horses. In the winter a sled will do.

If you are feeding hay to outside horses, have stacks of hay at convenient locations so you can walk through the paddock alleyways to feed. The larger the property, however, the more likely it is that you will need something larger to feed your horses. It is not fuel-efficient to use a tractor or truck to feed, but there are smaller, more fuel-efficient or even electric and battery-powered vehicles that are popular choices for feeding. We'll discuss these options in greater detail in chapter 10.

TRACTOR SHEDS

Store tractors and mechanical implements in sheds with at least three walls and a roof to protect them (and you, working on them) from the wind. You don't need heat or a door, but electricity is helpful for plugging in block heaters or powering lightbulbs. If tractor maintenance is required, you will be thankful for some shelter, as we all know that tractors break down in the most miserable weather.

Building a large enough shed is advisable with tractors as well. Because the tractor shed will have minimal power usage, you can build it a little bit bigger without being concerned with the environmental impact it will have. Therefore, build in extra room to allow for additional storage needs. You can always use miscellaneous storage space for the plethora of horse-related items we all tend to collect (winter-blanket storage in the summer, foaling gear, tools, and so on). Build your tractor shed to fit at least two vehicles. Depending on the size of your property, you may need a larger tractor for big jobs and a smaller one for dragging your arena or doing smaller jobs.

In northern climates this shed should have electricity running to it so that you can plug in your machinery. It may be entirely feasible to use a small solar power kit to power your outbuildings if only a single light or a plug-in for a tractor

is required. (Keep in mind that you should have a timer on your outlets to turn on just two hours before you need to use the vehicles. Any longer and you are just wasting energy.)

Other implements such as spreaders, tillers, or drags can be kept in covered areas that do not need electricity. For the sake of efficiency when maneuvering around your farm pulling implements, keep these near the tractor shed so you can hook up quickly and get to work. Sometimes having the most efficient setup is simply knowing where you can cut a minute or two out of your regular routine.

COMBINED STORAGE

It is possible to combine some items in storage; however, choose items that will not cross-contaminate. Tractors and fuel can be stored together, as can hay and bedding, but do avoid storing fuel and feed in the same building.

Some feed can also be stored in your barn or attached to the side of it. This is a popular option for the sake of efficiency and is viable as long as you are taking all appropriate fire-prevention measures, such as erecting a thick cement wall between hay storage and the barn. If a fire were to break out in the hay shed, the cement wall would hinder its progression into the barn. There are specific companies that can help create true fireproof barriers — your insurance company will have more information.

Planning Outdoor Horse Shelters

Horses that spend time outside need shelter from the elements. Your climatic region will determine the amount and type of shelter your horse requires. The most environmentally friendly shelter for outside horses is to use the protection of available tree stands, bushes, and hills. In many states and provinces, the law requires specific shelters, especially if a horse spends more than 24 hours outside. Some shelter options include:

Co-oping

If you live in a farming or equestrian community, you may want to consider co-oping, or sharing many group resources. This can mean anything from managing a large community hay shed that can receive exceptionally large loads of hay to sharing farm equipment such as tractors and implements.

TURNOUT

- Paddocks located flush to the barn may not require extra shelter, but you can construct a wall of flat boards perpendicular to the barn wall that provides shelter from two sides. Overhang from barn may help.
- Turnout paddocks that open into barn stalls are sufficient shelter if your horse has free access.

PADDOCKS

- Two walls and a roof provide sufficient shelter. Share walls with the neighboring paddock and use less wood.
- Freestanding shelters that are built to block prevailing winds and precipitation.

PASTURES

- Build shelters close to existing trees to maximize protection from elements.
- To help even the bottom-rung horse get space in the shelter, build a four-sided shelter in the shape of an X with a roof. Each side has two walls and a roof for shelter from four directions.

Regardless of the type of enclosure, maximize your use of your natural landforms and trees while conserving material use by building shelters with shared walls.

☼ ❄ CLIMATE VARIATIONS

It's important to understand the relationship between climate and land. In a sense, the land formations create the climate due to the elevation or the simple fact of latitude and longitude, but the climate influences the land as well in terms of what thrives and survives upon it.

Subarctic

Some areas of this zone have a much closer relationship to permafrost than to the rest of the continent. The permafrost can be continuous, sporadic, or isolated and means that any digging (for buildings or fencing) must take the frozen ground into consideration.

Humid Continental

Using natural land formations and trees will be very important when working within your climate. There can be a great variation in temperatures, and your building plans must accommodate this through water drainage, heating and cooling, and mud management.

Humid Oceanic

You have a higher level of humidity here than most other climate zones. There may be areas of your land that see more moisture, for greater periods of time, than others. While there is a high level of fertility in the soil in this area, it can be damaged when not properly cared for during the wet season.

Highlands

The rocky terrain in your area turns fencing into a big undertaking. Be sure to choose a fencing option that will last a long time — it may take a lot of time and effort to get the job done right the first time, but it's much harder to replace broken and damaged fencing every year because it has not been installed correctly.

Semiarid

Because you regularly experience a little bit of drought, you must take extra special care of the aboveground water and riparian areas. Protect them!

Arid

While the semiarid folks need to protect their water, you need to covet and hoard yours. Water is life in your zone; don't let any building plans damage or infringe upon the water on your land.

GETTING OFF ON THE RIGHT HOOF

Building green structures from scratch

A GREEN HORSEKEEPER not only takes care of the land for future generations but also considers the impact that any structure created today will have on its future inhabitants. In addition, we can aim to minimize our impact on the entire environment and rely as little as possible on the manufactured energy that we discussed in chapter 2. The way to achieve these goals is by building efficient structures that integrate themselves into the environment with as little impact as possible.

Keeping your horse in a healthy environment begins with his living space: the barn. In this chapter we'll examine the important aspects of building ecofriendly structures. Because your property is part of a greater system, we will also discuss the most common types of building materials and some alternative sources. There may be other structures on your farm, from the hay shed to equipment storage facilities, and when you build, you can make environmentally friendly choices for each by applying the same principles.

In this chapter we'll discuss building from scratch, so we'll assume that you do not have a barn already constructed. If there are already structures on your property, you'll find information on making them more ecofriendly in chapter 5.

WHAT MAKES A BARN GREEN?

Hint: The answer isn't a bucket of green paint — well, maybe if it's low-VOC paint. It's time to get into the nuts and bolts of green buildings. What better place to start than the barn, the central building on your farm.

Creating green structures requires knowing where your materials came from, how they were created, where they will go in the future, and what impact they'll have on the environment when they are no longer useful in their current state. These factors apply to the total anatomy of each building, taking into consideration the following:

STRUCTURAL ELEMENTS
- Foundation
- Roof
- Walls: interior, exterior, and internal

LEED BUILDINGS

The Leadership in Energy and Environmental Design (LEED) Green Building Rating System was developed by the United States Green Building Council (USGBC) to provide guidelines and standards for environmentally friendly building construction. Professionals can be certified as LEED Accredited Professionals (AP). In 2003 the Canada Green Building Council (CaGBC) also developed its own standards in a document called the LEED Canada-NC v1.0.

LEED rates buildings in six areas: sustainable sites, water efficiency, energy and atmosphere, materials and resources, indoor environmental quality, and innovation and design process. Buildings that meet certain prerequisites and earn credits in these six areas can qualify for four levels of certification:

- Certified: 26 to 32 points
- Silver: 33 to 38 points
- Gold: 39 to 51 points
- Platinum: 52 to 69 points

In addition, different rating systems apply to various types of projects:

- LEED for New Construction: New construction and major renovations (the most commonly applied-for LEED certification)
- LEED for Existing Buildings: Existing buildings seeking LEED certification
- LEED for Commercial Interiors: Commercial interior fit-outs by tenants
- LEED for Core and Shell: Core-and-shell projects (total building minus tenant fit-outs)
- LEED for Homes: Homes
- LEED for Neighborhood Development: Neighborhood development
- LEED for Schools: Recognizes the unique nature of the design and construction of K–12 schools
- LEED for Retail: Consists of two rating systems: one is based on New Construction and Major Renovations, version 2.2; the other track is based on LEED for Commercial Interiors, version 2.0

CIRCULATION

- Fuels
- Passive solar
- Evaporation
- Circulating air
- Connections
- Doors and windows
- Ducts
- Plumbing

As you begin to consider the various factors that make a green building green, also take into account low-impact construction, energy efficiency now and in the future, the life of the building, nontoxic building materials, and how the building integrates into the environment. In addition, you need to gauge the "embodied energy": the energy consumed in the materials' production, transportation, and installation. It may seem like a lot to think about, but it will soon become second nature to consider the past, present, and future of your buildings and their materials. It just takes a little intentional planning and a change in our customary mindset to expect all material producers to help us with the answers.

EVALUATING GREEN MATERIALS AND PRODUCTS

Just about every aspect of the anatomy of your barn can be evaluated for its shade of green, and you can determine how it fits into your vision of yourself as a green horsekeeper. With each product you consider, take into account the following questions:

- How much (embodied) energy was used to produce and transport the product and its components?
- What kinds of energy (renewable or otherwise) were used in producing it?
- What kind of pollution and waste did its production and transport generate, and how much will its disposal create?
- Can it be obtained locally?
- Does it make good use of a local resource, especially an overlooked or underused resource?
- Is it reused or recycled? If recycled, is it post-consumer (preferably) or postindustrial?
- Can it be recycled or reused at any point in its product life?
- How durable is it? How much and what kind of maintenance does it require over its lifetime?
- How well does it perform its system function, be it structural, climatic, surface, or other?
- If applicable, how does the material affect indoor air quality?

Don't be afraid to seek this information from retailers and manufacturers. If they cannot answer such questions, look up the material safety data sheet (MSDS) online (see Resources). All you need is the product identifier (such as the product code or style number) and the manufacturer. The MSDS sheets are regulated by the government and include information such as exposure situations, hazard prevention and protection, and other specific information.

Building Products Have Life Cycles Too

Every material used in construction has a life cycle. As you'll see in the following description, there is an obvious point where it begins and a point where its cycle ends, and during each stage energy is put into the cycle and a by-product is produced (very often carbon dioxide). That's why, when deciding if "green building products" are truly green, we must take this embodied energy into account. Embodied energy is essentially the sum of all the energy invested in the material from harvest through to implementation.

Products with the shortest life cycle (for instance, single-use items) are the most energy-inefficient because the ratio of energy used to produce the item and the length of time the item is in use are out of proportion. As we extend this cycle over longer periods of time by reusing and recycling, the item becomes more efficient.

This life cycle is broken down into three main phases:

- **Production.** The extraction of raw materials and the construction of the product
- **Use.** The transport, installation, and use of the product until its functional life span has ended
- **Return.** The point when the item is discarded or recycled

Each stage of this cycle can be deconstructed and evaluated for its environmental impact. This is where you can see the true investment of embodied energy in each stage. Some products — steel, for instance — will have much larger investments of energy before they have even reached your property to be installed.

PRODUCTION PROCESS

In the production process, the beginning of the product's life cycle, raw material is harvested from the earth, producing carbon dioxide during its extraction via machinery and transport to a

MAKING THE GRADE

So how do some common materials rank?
A = your best bet. *B* = not bad, when used correctly. *C* = doesn't quite pass the test.

YOUR BARN'S FOUNDATION

This is the part of your barn that will be intimately connected to the ground for a long time to come, so it must be as safe as possible. When managed correctly for its after-life, **cement** is a very ecologically friendly choice.

Q *Where does the material come from?*
A Concrete is made from a mixture of sand, gravel, or crushed stone; water; and Portland cement. Most of these products can come from recyclable material. *A+*

Q *What are the by-products of its creation?*
A It takes a lot of energy to make cement, including heating up to 2700°F (1482°C) and moving it with large trucks. *C−*

Q *How is it delivered and installed?*
A Mix the cement right at the job site. *A*

Q *How is it maintained and operated?*
A Once set, concrete is virtually maintenance-free and does not require any treatments, additives, or chemicals. *A*

Q *How healthy is this product?*
A There is no offgassing after installation. *A*

Q *How do we dispose of it?*
A While concrete can be recycled quite easily, it rarely is. Almost half the construction waste produced each year in the United States is from concrete and rubble. *B−*

Greenest Alternative

A concrete barn foundation is your best bet. Be sure to recycle it when the time comes.

ECOFRIENDLY STALL WALLS

The most common stall wall material has always been **wood.** Its sustainability depends on how you maintain it, how your horses use or abuse it, and what happens when you need to replace it.

Q *Where does the product come from?*
A Wood is usually harvested by mechanized equipment. *B* (or *B+* if the company invests in reforestation)

Q *What are the by-products of its creation?*
A The manufacture of wood products creates CO_2 emissions. *B*

Q *How is it delivered and installed?*
A Large trucks carry wood from forest to mill to store and finally to your farm. *C*

Q *How is it maintained and operated?*
A If a wood product is damaged, it is often replaced and the wood burned. There is little recycling of a broken board. *C*

Q *How healthy is this product?*
A Untreated wood is perfectly safe, but pressure-treated wood is unhealthy for humans and animals and cannot be recycled safely. *B*

Q *How do we dispose of it?*
A Pressure-treated wood must be taken to a landfill rather than recycled. For many years, laws in the United States mandated the use of arsenic-laced, pressure-treated wood for anything that came in contact with the ground or masonry, to prevent insect infestation and prolong the life of the wood. If the barn wood is old enough (and hence has not been treated), it may be recycled and sold. *C−* for pressure-treated wood and *A+* for recycled wood frames.

facility for manufacturing. This stage may also result in the destruction of natural resources.

During the manufacturing process, which converts raw materials into a specific product, unused materials are often discarded, recycled, or stored. Pollution created by factories is a green consideration.

USE OR FUNCTIONAL PHASE

The "use" stage of a product's life cycle begins with transportation and delivery of the product, which uses carbon dioxide as well as packaging for transport. Installation of the product may require power outputs and will result in discarded packaging and containers. Chemicals such as glue, sealants, or paint treatments may be applied.

The duration of a product's usefulness is unique to each product, including a stage of operation, when the product works as it should; a repair stage; and a point at which it breaks down and is removed, demolished, or destroyed.

RETURN OR END OF LIFE

Typically, most products are destroyed when they might be recycled or reused in a different way. Some products must be destroyed, such as arsenic-laced pressure-treated boards (see box, Making the Grade, on page 63); others can be recycled in some fashion.

Green Stall Floors

If your goal is to be completely environmentally friendly, it would be best for your horse to live outside for most of his life. Stalls require bedding, and straw and shavings both carry an environmental price tag because they must be harvested and transported using fossil fuels. In many areas of the country, however, and in many niche horse industries (competition, for example), it is often necessary to keep your horse in a stall.

Stall floors are typically made of dirt or concrete. While dirt can shift over time, causing sloping and uneven footing, concrete is too hard for horses to stand on for long periods of time. For covering concrete floors, rubber mats are a great choice because they use recycled rubber from tires or industrial use. They typically last five years in a stall that is used every day — a little bit less if the horse paws, pulls at the mat with his teeth, or has a high urine output. (Some horses prefer to urinate inside their stalls.)

Rubber mats create a cushion for the horse, but they are not water permeable, so the bedding layer soaks up the urine. Eventually, the rubber degrades and becomes thin and prone to tears. The pooling urine means that more bedding is soiled and has to be replaced each day.

WATER-PERMEABLE OPTIONS

One alternative is to install a water-permeable barrier that allows urine to drain away from the stall. This means less bedding is used to soak up the urine. Cleaning stalls will be a matter of cleaning out the manure and the small amount of soiled shavings.

Stall Skins, by Southwest Animal Products, are made from a nonwoven polypropylene material that allows urine to leach out through a water-permeable layer into the ground. This is another case where you must weigh the benefits of the material with the environmental costs associated with it. Polypropylene is used as a substitute for PVC because it emits less smoke and no toxic halogens when burned or thrown into a landfill. In fact, it is safe enough to be used in hernia-repair operations to protect the body from developing new hernias in the same location. However, there is no way to recycle this product into a usable material once it has worn out, increasing the likelihood that it takes up space in a landfill somewhere. In a case like this, where the product is safe and lasts a long time, you may be able to overlook the recycling issue.

GREEN ALTERNATIVES TO PRESSURE-TREATED LUMBER

Since you won't be using pressure-treated wood in your barn, what will you be using?

MATERIAL	DESCRIPTION	ADVANTAGES	DISADVANTAGES
Composites	Solid material made of wood fibers and recycled grocery bags and/or recycled milk jugs	Doesn't warp, split, chip, or rot; variety of colors to choose from; never needs sealing or staining; requires little maintenance; free of knots; resists moisture	More expensive; poor aesthetics; not rated for structural use (wood is required for structural supports and subframes); susceptible to mildew, mold, and stains; color fades in sunlight
Virgin vinyl	Hollow building material: a molecularly bonded blend of 100% virgin, high-polymer resin (some are made with UV inhibitors to prevent damage from sunlight and/or impact modifiers for greater strength)	Doesn't warp, split, chip, or rot; variety of colors to choose from; never needs sealing or staining	More expensive; poor aesthetics; not rated for structural use (wood is required for structural supports and subframes)
Redwood	Includes cedar and cypress	Resistant to decay and insects; good aesthetics; dimensional stability; never needs sealing or staining; easier to saw and nail	Expensive; surfaces are soft (susceptible to denting and scratching); susceptible to moisture
Exotic hardwood	Includes mahogany and a variety of ironwoods (IPÊ), also known as Pau Lope	Durable; resistant to decay and insects; never needs sealing or staining; good aesthetics; virtually knot-free; impervious to water	More expensive; difficult workability (requires predrilling for fasteners); environmental concerns (e.g., depleting supplies)
High-density polyethylene (HDPE)	Thermoplastic	Weather resistant; can be cut or drilled cleanly; no grain to split or chip; no need to predrill; suitable for agriculture uses; good low-temperature impact resistance; excellent chemical resistance	The EPA has not listed disadvantages
Rubber lumber	Composed of 50 percent plastic and 50 percent old tires	Durable; impervious to water; resistant to insects; resistant to UV rays; variety of colors; uses recycled materials	Color isn't guaranteed to last; not rated for structural use (wood is required for structural supports and subframes)
Tropical hardwood from sustainably managed forests	Natural tropical hardwoods originating from sustainable, managed forests, managed by the Forest Stewardship Council	Strong; durable, resistant to moisture; no changes in work practices or equipment; life cycle costs less than treated poles	Initially more expensive; supply not locally grown (factor in embodied energy and transportation costs)

Source: U.S. Environmental Protection Agency

Installing the skin is quite simple. Slope the stall floor slightly to create a leach area that should be filled first with gravel and then with sand to allow urine to leach away from the bedding. This approach also benefits indoor air quality (the ammonia smell is diminished) and your manure pile, which will have a higher concentration of manure in relation to bedding (more on this topic in chapter 7).

Hoof Grid is made of a durable, elastic high-pressure polyethylene that is resistant to fracture, UV, and frost. It is long-lasting, incorruptible, and nonpolluting, as well as resistant to ammonia, lactic acid, street salt, hot asphalt, and gasoline. And it's also 100 percent recyclable. It uses a "snowshoe principle," with grids that disperse the horse's weight over a larger surface area (effectively, 689 square inches [4,445 sq cm] for a four-legged animal).

Hoof Grid can be used in areas such as:

- Horse stables, paddocks, arenas
- Turnouts and riding paths
- Racetracks
- Side yards at homes
- Hillsides, for slope retention and erosion control
- Livestock wash areas
- Driveways and parking areas
- Drainage culverts and ditches
- Dog kennels and runs
- Grassy public-park areas
- Driving roadways through lawns

Hoof Grid can be quickly cut to fit at the installation site with an angle grinder or circular saw or for small jobs with a compass saw or handsaw.

To install Hoof Grid, the stall must have a layer of fine gravel over the foundation soil, with a slope of about 3 degrees, and then another layer of clean, washed, crushed rock over the top. Then lay the Hoof Grid sheets on the gravel and tamp down to connect the sheets. This setup allows urine to leach into the gravel so the nitrogen will dissipate and only the water will travel through the ground. (See Resources for more information.)

Hoof Grid provides support and drainage at the same time.

ECOFRIENDLY INDOOR ARENAS

Your indoor arena will benefit from good air circulation with minimal heat loss in the winter and good airflow in the summer. There are many different types of indoor arenas, from wooden structures that may include insulation to steel-framed, fabric-membrane buildings.

A Green Roof

As mentioned, a light-colored roof can make a great difference in the temperature inside your arena, but an even more environmentally friendly roof is a "green roof." Instead of using standard roofing materials such as asphalt shingles or metal roofing, you can build an "eco-roof," a roof that has been intentionally vegetated — in other words, you grow plants on your roof, on purpose. There are many reasons for considering this type of roof:

- Plants reduce CO_2 and produce oxygen.
- Heat is reduced inside the building during the day because the roof does not absorb and transfer thermal energy.
- The roof acts as an insulation barrier, even during winter months (though areas with heavy snowfalls are not necessarily good candidates for a green roof because of weight and structure issues).
- The roof provides a microhabitat for "good" bugs and birds that help maintain biodiversity in your area, resulting in a greater

number of native species that are able to survive and thrive.

- Plants will clean the air and prevent any polluting particles from entering the water system; essentially, they filter the air.
- The roof protects land from storm-water runoff, as it can absorb up to 80 percent of the water that falls on the roof. A study in Germany calculated that during a rainstorm when 211.3 quarts (200 L) of rainwater fell on an 18m2 green roof, only 15.9 quarts (15 L) fell from the roof to the ground.

Structural issues are always of the most concern, and you would be wise to consult an engineer to determine how to build a structure that can support a green roof. Germany, the country that leads the world in green roofs, has devised the following design loads for buildings through the German National Standard DIN 1055: Design Loading for Buildings (the load values are for saturated weights):

Gravel surface: 90–150 Kg/m² (0.13–0.21 psi)

Paving slabs: 160–220 Kg/m² (0.22–0.31 psi)

Vehicle surface: from 550 Kg/m² (0.77 psi)

Extensive green roof: 60–150 Kg/m² (0.08–0.21 psi)

Intensive green roof: 200–500 Kg/m² (0.28–0.7 psi)

A green roof needs a soil/gravel system that is engineered properly to capture enough water to sustain the plants and to provide adequate drainage, root barriers, and irrigation systems; and that is the correct weight-to-load-bearing-ability ratios. You can find a great deal of information about green roofs online (see Resources) or look in the bibliography for some books on this topic.

It is possible to construct a green roof on your own; however, if you feel that this project is not for you or not feasible, make sure that

Your Partners

There are 3,000 conservation districts in the United States. These government offices are established by each state to run natural-resource-management programs at the local level with the cooperation of landowners.

To find out which district you are in, look online: www.nacdnet.org/about/districts/directory/index.phtml.

Working with a local district has many advantages. They will be most familiar with your environment and its common pollutants, local regulations, financial incentive programs, and experts available for consultation or for hire.

Another service that you should become familiar with is the National Sustainable Agriculture Information Service (www.attra.org). According to its Web site, this organization, also called Appropriate Technology Transfer for Rural Areas (ATTRA), "is managed by the National Center for Appropriate Technology (NCAT) and is funded under a grant from the United States Department of Agriculture's Rural Business-Cooperative Service. It provides information and other technical assistance to farmers, ranchers, Extension agents, educators, and others involved in sustainable agriculture in the United States." Specifically geared to producers such as farmers, ranchers, market gardeners, and those involved in commercial agriculture, ATTRA also has services for those who are economically disadvantaged or who live in communities with few resources.

your arena roof is designed to help you collect rainwater. For every inch (2.5 cm) of rain that falls on a catchment area of 1,000 square feet (300 sq m), you can expect to collect approximately 600 gallons (2,200 L) of rainwater. We'll discuss this further in chapter 5.

HARNESSING THE SUN AND WIND

The sun plays a big role in having a comfortable, energy-efficient barn. Looking at your map, note how the sun travels over your property during the winter and the summer. Until the winter solstice in late December (in the Northern Hemisphere), the sun moves farther to the south, so windows or solar panels used for warmth should be placed on the southern face of a structure. In the southern half of the globe, they should be installed on the north side.

In the summer the sun is closer to the center of the sky while it travels from east to west (well, we all know it's the earth that is traveling, not the sun!). If you have windows in the summer that are used for cooling, you will want them on the eastern side of the building. The hottest part of the day in North America is the early afternoon, once the sun moves past its zenith (what many refer to as "high noon"). When the sun starts moving to the west, you don't want western windows letting more heat in, so that

end of the barn should be in the shade at this time of day.

Windows and vents can be used for heating *and* cooling, depending on your climate zone and time of year. To keep cool in the summer, close blinds or shutters but open the windows. To keep warm in the winter, open the blinds or shutters but close the windows. Any coverings should be located on the outside of the window, such as shutters, or attached to the outer wall, such as overhangs. Window coverings inside the barn can be hung only if the window is not located inside a stall (for the obvious reason that your horse may thoroughly enjoy himself while ripping down your handiwork).

SUNLIGHT AND SKYLIGHTS

Besides assisting with heating your property, the sun is the source of natural lighting. Keeping the sun from overheating a structure in the summer while allowing enough light in can be a tricky thing to balance.

Skylights are a great source of natural light, but often they are so high up that we have no way of closing them off when the sun is directly overhead and overheating the interior. Building a barn so that there is a roof over the skylight, or a tubular skylight that has the shaft offset from the opening so that more light than heat enters the building, is a much better idea.

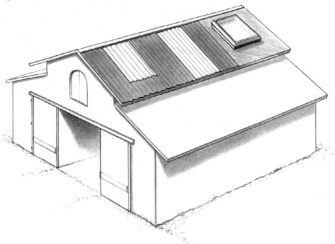

A barn can get too hot in summer with an overhead skylight, unless there is a mechanism to open and close it.

ARE SOLAR SHADES REALLY ENVIRONMENTALLY FRIENDLY?

Solar screens or shades — externally mounted, specially woven fabrics coated with polyvinyl chloride (PVC) stretched over aluminum frames and mounted on existing windows — may not be the environmentally friendly products that the manufacturers advertise. It is true that they block up to 90 percent of thermal heat and ultraviolet rays and that from the inside they are virtually unnoticeable, allowing light through the windows without heat. Groups like Greenpeace, however, have decried the use of PVC because of the by-products (dioxins) created during its manufacture.

The Technical and Scientific Advisory Committee of the U.S. Green Building Council (USGBC) released a report in 2007 for the LEED Green Building Rating system that concluded that "no single material shows up as the best across all the human health and environmental impact categories, nor as the worst," but that the "risk of dioxin emissions puts PVC consistently among the worst materials for human health impacts."

YOUR TEMPERATURE RANGE

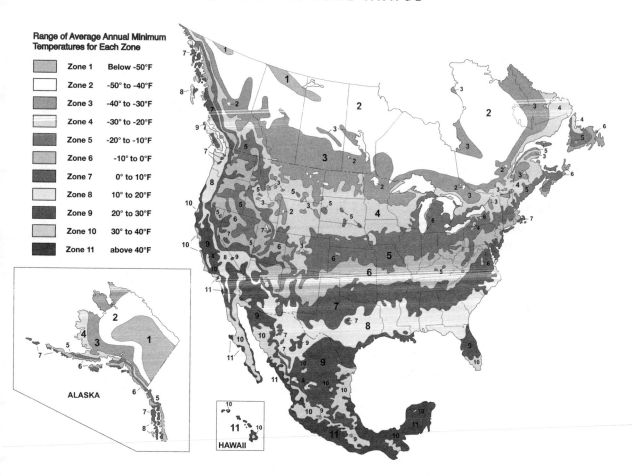

Range of Average Annual Minimum Temperatures for Each Zone

	Zone 1	Below -50°F
Zone 2	-50° to -40°F
Zone 3	-40° to -30°F
Zone 4	-30° to -20°F
Zone 5	-20° to -10°F
Zone 6	-10° to 0°F
Zone 7	0° to 10°F
Zone 8	10° to 20°F
Zone 9	20° to 30°F
Zone 10	30° to 40°F
Zone 11	above 40°F

ALASKA

HAWAII

Keeping Your Cool

Using air-conditioning in a barn is a foolish waste of energy when there are many healthier and more energy-efficient ways to keep cool in the summer.

The biggest challenge you will face when heating or cooling will be working with your region's climate. If you are in southern California, keeping cool may be your number-one concern. If you are in southern Alberta, you live in one of the few places on earth where the temperature has been known to fluctuate from 104 to minus 40°F (40 to minus 40°C) in a year.

As mentioned, if keeping cool year-round is your concern, the majority of your cooling windows should face north, as the south side of your building will receive the most sun. For the sake of airflow and lighting, however, you will need some windows on the south side of your structure. The south-facing windows should be under a roof overhang or installed overhangs that shade the entire window. You can also plant fast-growing, large shade trees to ensure good south-facing shade cover.

It's not likely that you are considering installing an air conditioner in your barn, and the good news is that you don't have to. But you should keep the temperature as low as possible during the dog days of summer (July and August), when soaring temperatures can lead to all sorts of problems for your animals, from dehydration to heat stroke.

A Penny Saved

A single 24-foot (7.3 m) PowerFoil fan uses, on average, only 440 watts of power. At $0.05/kWh that is less than $0.03 per hour to operate.

If you have designed your building to take advantage of prevailing winds and shady trees, then you've got a good start. Adding a large fan is the next step; more specifically, a high-volume, low-speed (HVLS) fan that can lower the temperature in your barn or arena up to 15°F (8°C).

Big Ass Fans is a manufacturer of large fans that can be installed in barns and arenas. The blades of these fans are similar to those on a helicopter: aerodynamically correct cast aluminum. They cut a swath that is about 24 feet (7.3 m) in diameter, and they move at a relatively slow speed.

A lot of thought must go into the installation of HVLS fans because they work best when maximizing the airflow to create a cooling breeze. This really makes a difference in barns where training or competitions are going on because often the work cannot stop for the heat. By reducing the heat stress, the show (or the work) can go on.

The glass in barn windows concentrates the heat, allowing heat in and keeping cooling air out, so creating shade over your windows greatly reduces that thermal-energy heating during the summer months. And yet you want that warmth in the winter months. To make full use of the heat coming in the windows in winter, install exterior shades or awnings over your windows that can be adjusted based on the amount of cover you require for each season.

Installing a light-colored roof is a great idea. Dark roofs can heat up to 180°F (82.2°C) on a hot, sunny day. While much of the heat is radiated back out, some is transferred into your arena or barn.

Staying Warm

HVLS fans work for heating as well, especially in barns and arenas that have high ceilings. Since you don't tend to spend much time up near the ceiling, you may not know that the temperature can be 20 to 30°F (11 to 17°C) warmer than the air on the ground. We all know that hot air rises; the HVLS fans mix that hot air back into the interior air, creating a higher average temperature. So if the air at your feet is 65°F (18.3°C) and the air at the top is 85°F (29.4°C), then an average temperature of 75°F (23.9°C) would result.

In barns and arenas that have high ceilings, you need to circulate the air three to four times per hour. This will allow you to set the thermostat lower while keeping the same level of comfort. According to the statistics for Big Ass Fans, for every Fahrenheit degree (0.6°C) the thermostat is lowered in the winter, you'll save about 3 percent a day in heating costs. Lowering the thermostat by 5°F (2.8°C) will bring you a 15 percent savings.

During the winter large fans like these should be set so the airflow is downward. This creates less "wind" or air velocity across the ceiling and therefore less heat transfer through the roof. By using a specific "heat-transfer coefficient" that calculates the amount of heat lost, Big Ass Fans has shown that if fans are reversed and air is being drawn upward, it results in 3.7 times more heat lost through the roof.

GREEN INSULATION

You can choose among several different types of insulation for the walls of your barn or arena. Insulation helps your building retain both warm and cold air, but unfortunately, many types of insulation use formaldehyde, a known carcinogen, as a binding agent. Happily, there are other options: formaldehyde-free recycled-cotton insulation; loose-fill cellulose made from recycled newsprint and treated to be fire resistant; or spray-in foams that are soy based.

Any type of insulation will require that you have interior walls to hold the insulation in, as well as moisture barriers (usually plastic sheets) to keep the insulation dry. Horses, like humans, release quite a bit of water as they breathe and perspire. This creates humidity inside your barn, and the barriers are there to keep the insulation from becoming damp and rotting. These barriers will also help keep pests and rodents from entering the walls.

Your roof can be insulated in the same way. If you have a loft to be used for storing hay,

---

Fan Facts

Large spaces like barns and arenas can be difficult to keep at a comfortable temperature in the winter. Using an HVLS can help stabilize the temperature, but only if they are operated correctly with airflow directed down and away from the ceiling. According to Richard Aynsley, Director, Research and Development, Big Ass Fan Company, on the company's Web site:

> If the airflow from the ceiling fan is reversed with airflow upward, even at low speed, the velocity of air across the ceiling above the fan is high, typically around 400 fpm. At this speed, the heat transfer coefficient at the ceiling is around 1.87 Btu/h.ft2.°F. However if the ceiling fan is running at low speed with airflow downward the airflow across the ceiling is low, typically 80 fpm. At this speed, the heat transfer coefficient at the ceiling is around 0.051 Btu/h.ft2.°F.
>
> In short, reversing the air flow direction from ceiling fans in winter means that heat loss through the ceiling is increased by a factor of around 3.7 times due to the increase in surface conductance.

insulation is very important to help maintain consistent temperatures. If heat and humidity become too high, your hay can spoil quickly, so you will need to keep your loft temperature consistent and provide some form of venting, especially in the summer.

If you will be building a metal barn or arena, ensure that you have some type of radiant barrier to stop heat from being transferred during the summer. Radiant barriers are basically aluminum-foil sheeting with a backing for stability and installation and are sometimes called "reflective insulation."

If you are going to maximize your thermal-energy transfer as noted in chapter 2, you may want to consider concrete blocks or masonry walls. You can still install reflective insulation on the barn interior to prevent drafts and provide condensation control. If you are located in the north, you can include a layer of rigid foam or spray foam between the wall and the radiant insulation.

VOCs and Outgassing

Commonly used in furniture, carpet, paint, and other building products, volatile organic compounds (VOCs) are toxic compounds — for example, formaldehyde — that leak out into the air and can cause damage to lungs and the environment. Offgassing or outgassing is the release or evaporation of chemicals (trapped, frozen, or absorbed) from materials that can be released directly into the air, or into soil or water, depending on what they are being released from and how it is being used.

Some examples of materials with VOCs are:

- Household products, including paints, paint strippers, and other solvents
- Wood preservatives
- Aerosol sprays
- Cleansers and disinfectants
- Moth repellents and air fresheners
- Stored fuels and automotive products
- Hobby supplies
- Dry-cleaned clothing

VOCs are used mainly in new construction but also can be used in remodeling or reconstruction. For example, plywood outgases formaldehyde, and many paints offgas a variety of chemicals, giving that "new paint" smell.

 R-VALUES

When searching for the proper insulation for your barn, pay attention to the "R-value." That is the unit of measure for the rate of heat flow through a given thickness of material. The higher the number, the better value the material has:

- ½" plywood = 0.02
- ½" gypsum wallboard = 0.45
- ⅝" gypsum wallboard = 0.56
- 8" concrete block = 1.11
- Insulated glass = 1.65
- Sheep's wool = 2.75
- Perlite = 3.0
- Cellulose insulation = 3.5
- Cotton = 3.5
- 1" mineral wool = 3.7
- 1" polystyrene insulation = 4.0
- 1" isocyanurate insulation = 7.5
- 3" polystyrene insulation = 12.0
- 3½" fiberglass insulation = 13.48
- 6" fiberglass insulation = 19.0
- 3" isocyanurate insulation = 22.5

GREEN INSULATION

TYPE	ABOUT	PROS	CONS
Cellulose	Made of newspaper, borates, and ammonium sulfate; cellulose insulation can be bound as a wet spray and installed by a professional in open wall cavities, resulting in entire walls that are effectively sealed.	R-value of 3–3.7; should contain a minimum of 75 percent postconsumer recycled content; has the lowest embodied energy of any insulation product because it can contain upward of 90 percent postconsumer recycled newspaper; comes in low-VOC variety	Manufacturing process does not cause significant pollution problems, but the fibers and chemicals used in cellulose insulation can be irritants; fire-retardant chemical additives such as ammonium, sulfate, boric acid, and sodium borate are used in cellulose insulation; small amounts of formaldehyde may be emitted from printer inks in recycled newspaper; main concerns are settling, displacement as a result of wind, and infestations of rodents
Cotton	Made of all-natural fibers; normally comes from postindustrial recycled cotton textiles, such as denim.	R-value of 3–3.7; should contain a minimum of 25 percent recycled content; formaldehyde-free and contains no carcinogenic respiratory irritations; can be recycled and requires very little energy to manufacture	None
Mineral wool	Combination of two different types of wool: Slag wool is an industrial waste product produced from iron-ore blast-furnace slag. Rock wool is produced from natural rocks of basalt and diabase and provides great energy performance; will not burn and is chemically inert.	R-value of approximately 3.1–3.7; diverts waste from landfills because it's made from waste materials; it's recyclable, durable, fireproof, and naturally resistant to rot	Bacterial growth is a problem if it becomes wet; can emit toxic fumes when burned; not bio-degradable; can cause irritation to the skin, eyes, and throat, so a dust mask, goggles, or glasses should be used during installation
Perlite	Ethylene propylene diene monomer (EPDM) rubber; common type of low-slope roofing material in the U.S.	R-value of 2.5–3.3; very low thermal conductivity; high fire resistance, large surface area, and low moisture retention	Dust created may be an irritant
Sheep's wool	Mary had a little lamb . . .	R-value of 2.0–3.5; natural material; biodegradable and recyclable, very little energy used to produce; naturally resistant to insects and decay; inherently moisture tolerant (without losing thermal efficiency); can soak up and neutralize harmful toxins such as nitrogen oxide	Must be protected from water leakage; if the borate treatment fails, the material will lose its fire and mold resistance; can be attacked by moths if untreated

LONDON 2012

During London's bid for the 2012 Olympics, the organizing committee crafted its proposal around the concept "Towards a One Planet Olympics." They focused on five key areas: combating climate change, reducing waste, enhancing biodiversity, promoting inclusion, and improving healthy living.

According to the committee report, they are "beating the target to reclaim 90 percent of demolition material for recycling or reuse; reclaiming materials to reuse in designs of venues and parklands; recycling complete buildings to be re-assembled off site; and translocating wildlife and creating new habitats including a wildlife corridor to the north of the Park."

The ambitious reclamation project will create aesthetic and practical features for the park, including paths, paving and paving inlays, benches, planters and lighting, and water features. The total amount reclaimed as of the summer of 2008 has exceeded 90 percent of demolished materials, including the following: 80 lampposts; 160 manhole covers and 187 gulleys; 18 square meters of clay and slate roof tiles; 2 tonnes of red bricks; 117 tonnes of Yorkstone; 100 tonnes of cobble/granite; 41 tonnes of paving bricks and 35 tonnes of paving slabs; 1,200 m of granite kerbs and 4,200 m of concrete kerbs.

They follow their five focus areas to the nth degree by designing energy-efficient buildings, using lower carbon alternatives, building a wind turbine, and selecting materials that require less energy to produce. In addition, they have transported (or plan to transport) half of the construction materials by water or rail, thereby reducing greenhouse-gas outputs.

The committee described the existing site: Much of the site itself is contaminated, derelict and abandoned. The waterways in the area have suffered from years of neglect: water quality is poor, river walls are in a bad condition and the landscape is scarred with rubbish strewn along the river channels. For the past 400 years much of the area has been used for industry, from textile printing in the seventeenth century to petrol factories in the eighteenth century. Bone, varnish, soap, and tallow works, along with distilleries, engineering and chemical plants, have all been located in the area in the past. Since the late nineteenth century around half of the site has been used for landfill, including a 100-year-old tip on the site of Velo-Park. About 75 percent of the land has some form of contamination, such as petrol, oil, tar and heavy metals, such as arsenic and lead. The waterways that criss-cross the Park have suffered from years of neglect and the skyline is dominated by the 52 pylons that carry power lines across the area.

So how did they clean up this dirty, neglected area? First they removed 2 million cubic yards (1.5 million cu m) of soil and cleaned it with "soil-washing" machines that wash, sieve, and shake out pollutants like petrol, oil, tar, arsenic, and lead. The soil-treatment center and three soil-washing machines can wash 220 tons (200 metric tonnes) of soil per hour.

Invasive weeds were removed through incineration, and friendly flora and fauna native to the area were collected and relocated so they can become part of the postconstruction landscape. Areas of plant growth near the water are being protected from construction and will be incorporated into the design.

This is just one large-scale example of the possibilities for green, sustainable projects despite seemingly insurmountable odds.

Some climate considerations for building green in different regions follow.

Subarctic

Since you are more likely to be battling the cold than the heat in the subarctic climate zone, you will need to focus on buildings that utilize as much thermal energy as possible. Insulating your arena and building barns that are smaller are very important strategies to utilize as much heat as possible while minimizing heat loss through leak management. Most likely your outdoor arena will be used for only a couple of months out of the year, so invest most of your money in your indoor arena.

Humid Continental

Well, aren't you in a bit of a catch-22? In the humid continental zone, you have such crazy weather patterns some days that the only thing you can be sure of is "precipitation and lots of it." In the winter you see a lot of snowfall, especially if you live near the Great Lakes and "benefit" from lake-effect snow. In fact, those of you below the 49th parallel receive more snow than some parts of Canada. If you are on the East Coast, your weather is a little more temperate, with less snow but more rain.

Long story short, you need to make sure you are managing your water flow for rainy seasons and for spring thaws. While the high levels of precipitation might lead you to believe that green roofing isn't for you, I suggest you speak to an engineer before you make a final decision.

Humid Oceanic

Most of the rest of the continent thinks you folks in the humid oceanic region have it too easy. You do, however, have your challenges. Sometimes it is warm and humid but no rain is falling. This makes things difficult because you could *use* that rain if it would just *fall* rather than stay suspended in the air as water vapor. Because heat and humidity may be a concern, you should investigate some shade options for keeping your horses cool outside. Drought is rare in your area, so you should also learn where your water is flowing and how to use it to your benefit.

Highlands

Spring melt can be a dangerous time in the highlands area, especially if you are in a zone (Alberta or Montana, for instance) that experiences Chinooks that cause quick snowmelts. One minute you are calf deep in snow, and the next you've turned the corner into spring, and it's all melting. Managing runoff from mountain ranges can be tricky: the ground can be saturated easily in a short time because of the rockier terrain (there is less soil to retain water).

Make sure you are building shelters for wind and shade cover. If you are in the foothills, you may be able to take advantage of trees for this.

Semiarid

If you live in the semiarid climate zone, make sure that you are using your pastureland correctly, and include sacrifice areas in your plan. Choose plants that conserve water, and manage the water runoff from your roof as best you can for irrigation and watering horses. If you choose a green roof, you may have to spend more time watering it, so choose natural grasses that survive on less water.

Arid

Did you want heat? Well, if you live in the arid zone, you have plenty of it. Your indoor arena should be shaded. Include fans to get that air moving; otherwise, it will be unusable for a good portion of the year. Use all water runoff from buildings, and conserve water as much as possible. Pray for more rain.

REBUILDING AND REFITTING

Go greener by adapting existing structures

SINCE THE AGE OF INDUSTRIALISM BEGAN, our technologies have been advancing at a steady pace. Someone is always out there devising a better way to build something bigger and quicker. Globalization means that pieces and parts of buildings may be manufactured thousands of miles away, brought to us by ship, truck, and rail, and assembled on our property.

A benefit of these decades of growth and consumption is that many structures are in place that have long since cashed the environmental-cost check and now just need to be used properly.

WORKING WITH WHAT YOU HAVE

If you think you already might be behind the eight ball because you are not able to start an environmentally friendly horse operation from scratch, you are wrong. In fact, it is better for the planet when you use the facility you have and make a few adjustments, rather than start new construction projects.

As we've discussed, when you start brand-new construction, you increase the output of carbon dioxide into the atmosphere, because each new item that finds its way onto your property has been harvested, packaged, and transported to you. If your farm already has structures, most of the basics you need are right in front of you. All it takes to be more ecofriendly is a little time spent learning about how best to adapt green practices to your property and upgrading to some green products and processes.

Does LEED Apply?

One common criticism of LEED (Leadership in Energy and Environmental Design Green Building Rating System, see page 61) is that its main focus is on new construction. As mentioned, refurbishing older structures to be more energy efficient will always be more environmentally friendly than building an entirely new building.

Restoring Damaged Land

A lot of damage can be caused when buildings and roads were not located properly on your property. They may have been placed where they are for the sake of convenience or because the property was designed with a different purpose in mind. What once was farmland with fields and a homestead might now be divided into many 10-acre parcels with concentrated horse farms or be home to a breeding and training operation.

First you will need to identify damaged areas on your property. These might include roads that flood every spring, banks and hills where the grass is no longer growing due to erosion of topsoil, or paddocks that are no longer able to support natural forage and grass. Several of these topics will be covered later in the book, but for now you just need to understand the importance of evaluating your property as a whole system.

Even the simplest thing, like water that pools too long, can lead to larger problems (like mosquito infestations) that might tempt you to use chemicals to eradicate them. Certainly, in a time when West Nile virus is a concern, we may be more apt to overlook environmental considerations and instead use whatever chemicals are necessary to get rid of the disease-carrying insects. But a much healthier option is to use green landscaping techniques that direct water to nourish crops and pastureland and away from low-lying areas where it can play host to the newest generation of mosquitoes.

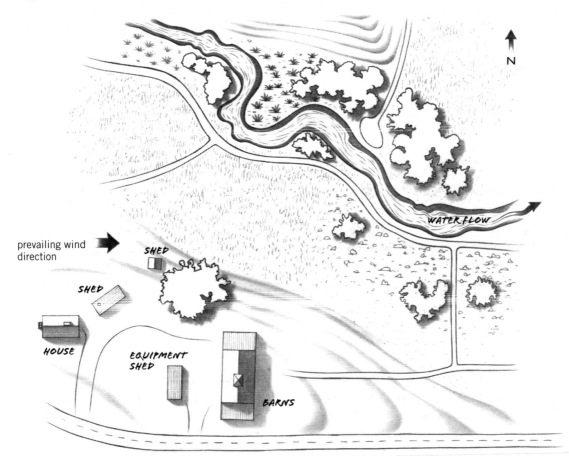

This is how your map might look with existing buildings.

In this chapter we'll discuss the following aspects of making your property more environmentally friendly:

- Efficiency
- Nontoxicity
- Introducing green products

 Taking Inventory

In a notebook, start a catalog of all the buildings on your property that you currently use. These may include some or all of the following:

- Barn(s)
- Indoor arena(s)
- Hay shed
- Tractor shed
- Garage
- House or apartment

Making Old Buildings Green

Your property may be home to several buildings in various states of repair and use. Some may be newer, not yet in need of major repairs. Some may be less important, to be fixed up "when you get around to it," and some you might be working on right now in your "spare time" or when you're not reading this book.

As mentioned above, reusing older buildings (for the same or different purpose) and making them more efficient are among the greenest actions you can take. The majority of the energy inputs and their resulting by-products have long since been invested because these environmental costs are front-loaded in the first few stages of the life cycle of your building. (See chapter 4 for more on the life cycles of building materials.) Converting an old chicken coop into a tractor shed or a foaling barn will always be more environmentally friendly than tearing one building down to erect another in its place.

LIMITING METAL USE

According to Eric Corey Freed in *Green Building and Remodeling for Dummies*, metal is an attractive choice for builders because it is durable and easily recyclable. But what Freed notes and what many fail to take into consideration are the incredible amounts of energy that go into creating metal in all stages of production. First it must be mined, then melted down at high temperatures, which releases a great deal of greenhouse gases.

The type of metal you use may also have other significant impacts. Copper is often mined in South America under abhorrent working conditions and using slave labor. Mining bauxite to make aluminum causes

excess pollution, and the creation of stainless steel can produce chromium compounds that are highly polluting and toxic.

Simple economics have led many manufacturers to save costs by including some amount of recycled content in the metal. If you must use metal, look for the product that has the highest amount available. Don't forget to investigate products made of 100 percent recycled material, often in salvage yards.

And as a final note, remember to never mix metal outdoors. A copper roof must always be installed with copper nails and a zinc roof with zinc nails to prevent a chemical reaction from occurring. (They swap ions in a complicated exchange.)

BARN VENTILATION

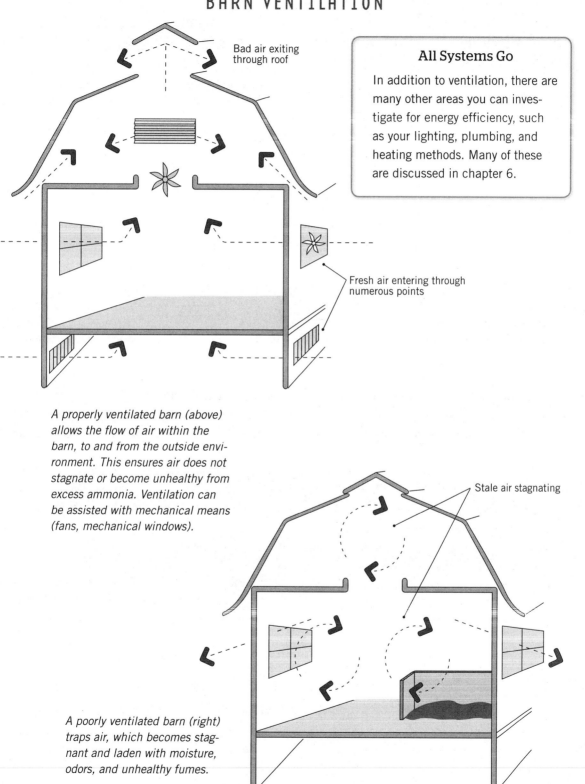

Bad air exiting through roof

All Systems Go

In addition to ventilation, there are many other areas you can investigate for energy efficiency, such as your lighting, plumbing, and heating methods. Many of these are discussed in chapter 6.

Fresh air entering through numerous points

A properly ventilated barn (above) allows the flow of air within the barn, to and from the outside environment. This ensures air does not stagnate or become unhealthy from excess ammonia. Ventilation can be assisted with mechanical means (fans, mechanical windows).

Stale air stagnating

A poorly ventilated barn (right) traps air, which becomes stagnant and laden with moisture, odors, and unhealthy fumes.

THE BARN: A BALANCING ACT

A barn is a unique structure. It must balance ventilation with energy efficiency: get the bad air out, let the good air in, and keep it warm or cool, as needed. For barns in Canada and the northern states, keeping warm in winter is an essential aspect of a well-built barn. We cannot sacrifice good ventilation for warmth, however; indoor air quality in barns is very important for the health of horses and humans.

Look Before You Leak

Make sure all leaks are sealed so you can direct the air where you want it to go while maintaining your ideal temperature. To do this, you need to go on a leak hunt. Heat is typically lost through two sources: windows and doors. Surprising, isn't it? More specifically, heat is lost through windows and doors that are not correctly sealed. If you were to hold a match (if you try this, be careful — you are in a barn) near unsealed, closed windows and doors in your barn, you would easily detect air leaks. You can test the seal on doors by shutting a piece of

---

Gather Your Tools

For repairing leaks you will need:

- Clear siliconized acrylic latex caulk
- Caulking gun
- Filler caulk for larger holes
- Weather stripping for doors and windows
- Insulating gaskets for electrical outlets

Holes and gaps larger than ¼ inch (0.6 cm) can be filled with expanding foam, duct tape, and so on.

Sealing Leaks in Ductwork

The farther north you live, the more likely it is that you will have a heater and ductwork inside your barn. You can have all ducts examined, diagnosed for leaks, and sealed using the Aeroseal method recommended by Energy Star. The U.S. Department of Energy and the Electric Power Research Institute have released information stating that from 25 percent to 40 percent of the heating and cooling energy put out by heating, ventilating, and air-conditioning systems is lost through the duct system. The Department of Energy awarded Aeroseal the "Energy 100" Award for its duct-sealing technology.

---

paper in the door. If you can pull the paper out, you are losing energy through that door.

Filling the Gaps

Modern doors and windows are constructed with weather stripping that needs maintenance every year, especially in a barn, where there is more dirt and activity than inside your home. During your monthly maintenance checks, you should look over the stripping to see if it needs to be replaced. If the stripping is worn out, leaving a gap just wide enough for a pencil to fit between it and the door, you might as well have a 2-inch-square (13 sq cm) hole in your wall, because that's how much air is getting through.

WINDOWS

All visible cracks and gaps between nonmoving parts of the window trim should be sealed with weather stripping and caulk. If windowpanes are loose, use putty to stabilize them. This may be a large job, especially if windows are high up in the barn.

DOORS

Maintaining a correctly sealed livestock entrance can be quite difficult. These doors take quite a bit of abuse in their lifetime from being banged and bumped, slammed, and left open. The important thing here is to maintain the level of the ground around the door so it can open and close safely, without jostling or pulling. Forcing a horse to enter and exit through a door that is too small for it is not only unsafe for the horse but also risks damage to the doorway. A small tear in the weather stripping from a horseshoe can cause a great deal of heat loss over time.

WALL FIXTURES

Light switches and electrical-outlet covers should be insulated with foam gaskets behind them. These seal the holes created when the switches are installed. Any unused outlets should have child safety plugs inserted to prevent cold air from coming in. Also seal up any spot where something has been cut into the wall. This might range from light fixtures and plumbing pipes to automatic waterers, toilets, sinks, and faucets in the wash rack.

THE HAYLOFT

If you have a hayloft above your barn, make sure the floor of the loft is sealed to prevent warm air and humidity from escaping up into it — your hay doesn't need to stay warm in the winter.

GREEN BUILDING PRODUCTS FOR BARNS

MATERIAL	PROS	CONS	ALTERNATIVE
Granite and stone	Harder stones are resistant to environmental damage through wind, exposure, chemicals, and so on; no maintenance or upkeep necessary	Nonrenewable; extraction and transportation carry high environmental costs; softer stones (marble, sandstone, limestone, slate) scratch easily and absorb stains	Salvaged stone at local salvage yards or "boneyards" at stone or marble yards
Paper resin, made from paper and a resin binder; often used in countertops; only one-third plastic	Hard and dense; good for countertops in barn office and bathroom; some companies use recycled paper	None	Can replace laminate countertops or cupboards in office, tackroom, bathroom
Aluminum siding	Low maintenance; easily recycled	Difficult to find a source for new siding made of recycled content, as aluminum siding has been replaced on the market by vinyl siding (which you won't want to use due to the environmental cost)	Wood siding certified by the Forest Stewardship Council (FSC) to be sustainably harvested
Wood	Relatively inexpensive; attractive	High environmental cost if harvested from old-growth trees, such as redwood	Wood certified by the Forest Stewardship Council (FSC) as sustainably harvested; composite lumber (combined recycled plastic and sawdust); IPÊ wood (strong; natural water resistance)
Asphalt shingles	Inexpensive	You cannot use the water you collect off asphalt shingles because of the oil in the asphalt	Recycled rubber shingles (best in northern climates, where it's okay to increase heat in the barn); recycled plastic and metal shingles; spray-on foam roofing (polyisocyanurate is great for flat roofs and is very light in color)

This fix can be as simple as laying a heavy-duty plastic sheet down on the floor of your loft and using 2×4 boards to secure it to the wall. The plastic goes between the board and the wall and is held securely in place, creating an air barrier between the loft and your barn. Sealing from the inside out is essential. If you seal up your barn from the outside, moisture can be trapped, causing condensation inside your walls and creating the perfect environment for mold to form in your hay, which may lead to sick horses.

PRESSURE-TREATED WOOD

While refitting your barn and property with greener materials and for more efficient energy uses, you may encounter existing pressure-treated wood. According to the Massachusetts Department of Environmental Protection, "You can usually recognize pressure-treated wood by its greenish tint, especially on the cut end, and staple-sized slits that line the wood. However, the greenish tint fades with time, and not all pressure-treated wood has the slits. If you are uncertain what your structure was made of, try contacting the manufacturer or builder."

If you are unable to remove the wood right away, coat it with at least two coats of water-based sealer and reapply every two years.

The process of pressure-treating wood, which has been around for about 75 years, involves infusing wood with preservatives to create a product that will not rot or decay for more than 20 years. This was deemed a great advancement because fewer trees had to be harvested. In fact, Wood Preservation Canada still touts pressure-treated wood as the way to "conserve our forest resource by ensuring that wood lasts longer in situations where termite and fungal decay hazards are high."

When wood is pressure treated, it is placed in a large tank that is then depressurized (all air is removed). Chemical preservatives are pumped into the tank under high pressure, forcing them deep into the wood.

Three primary chemicals are used:

- Creosote: oil-borne preservative used mainly for railway ties, mine timbers,

---

REMOVING PRESSURE-TREATED WOOD

The greatest threat is not wood that you may purchase (we will include many alternatives here) but wood that may already be on your property. Even old treated wood presents a hazard when it is cut and the dust is inhaled or ingested.

No pressure-treated wood can be recycled safely and must be taken to a landfill; nor should it be burned. When removing the wood, wear gloves, a long-sleeved shirt, and long pants. Wear a dust mask to avoid inhaling of any of the preservative-laden particulates.

Wear safety glasses or goggles, and wash work clothes separately from other laundry to avoid cross contamination.

Some states have laws regarding the disposal of this wood. For example, in California all pressure-treated lumber must be disposed of in a hazardous-material landfill or in a composite-lined portion of a municipal landfill that meets specific requirements. Contact your local landfill to inquire about your state or province's legislation.

poles, foundation piles, marine piles, and bulkheads

- Pentachlorophenol: oil-borne preservative used on utility poles, cross arms, and bridge timbers and ties
- Chromated copper arsenate (CCA): water-borne and used in guardrail posts, utility poles, bridge timbers, piles, structural glued-laminated timbers, posts, and permanent wood foundations

By far the most common preservative is CCA, which is related to arsenic. Wood Preservation Canada justifies this use "because of the differences between the arsenic compounds. CCA treating solutions are made with hexavalent chromium, cupric oxide, arsenic pentoxide and water with the solution concentrations ranging from 2 percent to 4 percent, depending on the species of wood being treated and the end use.

"The primary arsenic compound used in CCA is inorganic pentavalent arsenate, a naturally occurring trace element which is found even in human tissue and rapidly excreted by the kidneys without accumulating. It is quite different than trivalent arsenic compounds that are more toxic and often associated with chronic inorganic arsenic intoxication."

The group's defense notwithstanding, just because a substance occurs naturally does not mean that it is healthy to ingest. The compound is water soluble and may leach (or "dislodge") when exposed to water. This is a concern because horses chew wood and salivate and may ingest small quantities of CCA.

CCA was investigated by the U.S. Environmental Protection Agency (EPA) in 2002, and an agreement was made with the wood industry to voluntarily discontinue using CCA-treated wood for residential use (including children's playgrounds). Health Canada's Pest Management Regulatory Agency made the same announcement later in the same year. Both countries, however, allow its use for industrial projects, so it may be used in barns and fences.

Greener Alternatives

For residential lumber, two wood-treatment alternatives are promoted: amine copper quat (ACQ) and copper azole (CA). Both of these chemicals, however, are quite a bit more expensive. Different levels of the chemical are infused depending on whether the lumber is to be used above ground or in it. The more popular ACQ was created by Chemical Specialties, Inc. (CSI, now Viance), and the company received the EPA's Presidential Green Chemistry Challenge Award in 2002 for commercial introduction of ACQ. Its widespread use has eliminated major quantities of arsenic and chromium previously contained in CCA.

The high levels of copper in both ACQ and CA, however, mean that they are more corrosive to steel, according to American Wood Preservers test results, and therefore you must use double-galvanized or stainless-steel fasteners in ACQ timber. For additional information visit the U.S. Forest Service's Forest Products Laboratory Web site or learn about non-CCA wood alternatives at the National Pesticide Information Center's Web site (see Resources).

If you are seeking yet another alternative for pressure-treated wood, look for borate pressure-treated wood that uses EnviroSafe Plus, a colorless, water-repellant polymer (see Resources for Web site). This lumber can be used in indoor and outdoor construction.

COLLECTING RAINWATER

If you are not already collecting rainwater from your barn or arena roof, you are missing out on a great resource! Water collected can be used for watering horses, keeping dust down in arenas, and hydrating gardens and fields. (Colorado residents, see box on page 86.)

MAKE YOUR OWN RAIN BARREL

Using gutters, downspouts, and water barrels is a simple way to catch water for immediate use. The City of Bremerton, Washington, offers the following directions for making your own rain barrel (see Resources). Colorado residents, see note on page 86.

TOOLS NEEDED

6" hole saw (a saber saw, a keyhole saw, or
  a drywall saw will also work)
Drill
$^{29}\!/_{32}$" drill bit
¾" pipe tap

MATERIALS NEEDED

55-gallon food-grade-quality recycled barrel
Louvered screen
¾" brass faucet
Teflon tape or all-purpose caulk
¾" hose adapter

STEP 1: PREPARE THE BARREL

- Use a 6" hole saw, a saber saw, a keyhole saw or a drywall saw to cut a perfectly round 6" hole on the top of your barrel.
- Drill two holes with a $^{29}\!/_{32}$" drill bit, one toward the top for an overflow and one toward the bottom of the barrel for the faucet.
- Next use a ¾" NPT pipe tap and twist it into the upper $^{29}\!/_{32}$" hole, then untwist the tap and back it out of the hole, then repeat the same process for the lower $^{29}\!/_{32}$" hole.
- Rinse your barrel out thoroughly, as it previously had a food product in it. Avoid using bleach, as it is environmentally harmful in the storm drains. For an environmentally safe soap solution use 2 teaspoons of castile soap and 2 teaspoons of vinegar or lemon juice for every gallon of water used to clean your barrel.

STEP 2: ATTACH THE FAUCET

- Twist the threaded side of the hose adapter into the ¾" threaded hole toward the top of the barrel.
- Prepare the threaded side of the brass faucet by either wrapping it tightly with Teflon tape, making four or five rotations until all the threads are covered, or applying a thin ribbon of Kitchen and Bath All Purpose Adhesive Caulk, or similar sealant.
- Twist in the threaded and now prepared end of the faucet into the ¾" threaded hole toward the bottom of the barrel.

STEP 3: PUT SYSTEM IN PLACE

- Cover the 6" hole in the top by placing the 6" louvered screen onto the barrel with the louvered side up and the screen side down.
- Slide a hose onto the hose adapter at the top of barrel to direct the overflow water away from your home.
- Place two cinder blocks under the selected downspout and place the barrel on this raised base.
- Cut your downspout about 4" above the top of the barrel, add an elbow, and make any final adjustments to the base and barrel.
- Add a hose on the faucet or keep it available to fill a watering can.

The benefits of rainwater collection include:
- Free of charge
- No need to transport or use complex and costly distribution systems
- Can supplement or replace groundwater supplies if groundwater is unavailable
- No need for water softeners
- Sodium free
- Superior for landscape irrigation
- Reduces need to divert rainwater on the ground
- Demand on current water utility reduced along with your utility bill

Permanent Cisterns

You can purchase rain barrels already set up with taps and hoses in sizes that range from 50 to 75 gallons (189–283 L), but if you are really serious about using rainwater, you may consider setting up a larger, permanent cistern. Using cisterns for water collection is actually an ancient technology employed by many cultures, such as the Romans and Aztecs. Beneath the city of Jerusalem rests a cistern that can hold 2 million gallons (7,570,824 L). Cisterns can be made of almost any material, but a great choice is ferrocement, a very tough material when it has hardened and both fire- and earthquakeproof. It's used frequently in the repair of ships, which indicates how waterproof it is. Ferrocement is available at most masonry-supply stores.

Constructing such a cistern can be a labor-intensive project, as ferrocement must be kept moist during hardening and may take up to a month to fully cure before it's ready to be used. Other options include cement, plastic, fiberglass, metal, and redwood, but ferrocement and cement are your two most environmentally

---

DEFINITION: Ferrocement

A composite material made with cement, sand, water, and wire or mesh material.

---

friendly options. Cement should be sealed with a product such as Damtite Powder Foundation Waterproofer to ensure it does not leak.

Choosing the location of a cistern is vitally important — water is heavy. If you are undertaking this project at the same time as building a barn, you can actually build it at the same time as you do your foundation and connect the two, ensuring stability. Otherwise you can choose either aboveground or underground cisterns. A cistern located underground requires less structural reinforcement because it will be surrounded by earth. Whether you choose above- or belowground, be sure to place the cistern on a solid pad, either cement or compacted, hard ground. It should not be excavated ground, as that has a tendency to settle with time.

Six Ways to Capture Water

There are six basic components to a water-catchment system:

1. CATCHMENT SURFACE (ROOF)

Metal roofs make the ideal catchment surface, though cement and wood will work as well. Other materials may include terra-cotta clay tiles, fiber-cement shingles, spray-on foam, slate, and recycled plastic. It is most important that you ensure there have been no chemical additives to the shingles or roof tiles, as there are with asphalt shingles. The smoother the surface the better, because the longer it takes the water to reach the cistern, the more that can evaporate.

2. GUTTERS AND DOWNSPOUTS

Gutters and downspouts are the part of the water-catchment system that channel water from the roof to the tank. Seamless aluminum gutters are the best choice because they lack solder, which may contain lead, at the joints. However, they often require professional installation and may be more expensive. The gutter slope around your entire roof should be toward the downspout to facilitate quick water flow. You may need to consult with a gutter expert if your roof has one or more valleys; a valley is that point where two roof peaks meet in the middle. Having water that flows in different directions to the downspout can be difficult to account for when planning.

3. SCREENS

There are several types of screens to consider: screens that remove debris such as leaf screens, first-flush diverters, and roof watchers. These should be easily accessible for cleaning and be located at various points along the gutter and spout system. If debris is left to gather, it's a logical assumption that things will become clogged. There is also a bacteria risk if damp leaves are left sitting for too long inside the gutters or spouts. You can experiment with

different types of screens, from steel mesh to screens made from nylons (pantyhose) over spouts. Flush diverters do not allow the first few gallons or liters of water to enter the cistern: they are often the dirtiest, having just washed the roof clean.

> ### Warning for Colorado Residents!
>
> Did you know that rain barrels are illegal in Colorado?
>
> Colorado water law requires that precipitation fall to the ground and run off and into the river of the watershed where it fell. Because rights to water are legally allocated in this state, an individual may not capture and use water to which that individual does not have a right.

A RAIN CATCHMENT SYSTEM

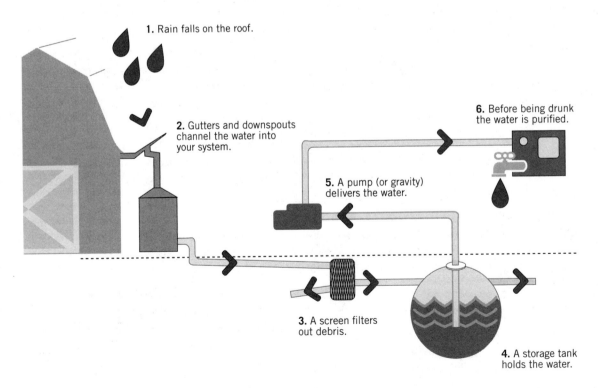

1. Rain falls on the roof.

2. Gutters and downspouts channel the water into your system.

3. A screen filters out debris.

4. A storage tank holds the water.

5. A pump (or gravity) delivers the water.

6. Before being drunk the water is purified.

4. CISTERNS OR STORAGE TANKS

These tanks should be opaque to inhibit algae growth, be completely clean, be covered to prevent mosquito breeding, and be accessible for periodic cleaning. They are the most expensive component and should be located as close as possible to the end-use location. A concrete pad is an excellent choice to place the tank on.

5. DELIVERY SYSTEM

A delivery system can be either gravity fed or pumped, and the type you choose will depend on the size of the tank. If you are implementing a smaller system that has smaller, aboveground barrels or catchments, you can use a very simple gravity-fed system, as described earlier. These systems are quite economical and typically work best if you receive at least 9.8 inches (250 mm) of water a year, which most of North America does (with the exception of some parts of the southwestern United States).

Placing barrels at all four corners of your barn and arena will allow you to make the best use of the water without having to drag hoses all across your property. Since these smaller systems have a limited supply (only as much as the barrel can hold), you don't want to use up the water by having it sit in a long hose. Once the water has left the barrel, it may not travel all the way to the end of the hose; if you need more pressure, simply raise the barrel higher by placing it on some cinder blocks. Make sure you roll up your hose starting at the barrel and drag the hose in to allow the water to vacate — this will prevent water from sitting in the hose and stagnating, which may encourage bacteria growth.

When using a barrel-and-tap system, make sure that the water is being filtered first to prevent the line from becoming clogged. This can be accomplished by using a simple grate or mesh to catch leaves. This needs to be cleaned out weekly.

Your Garden Will Thank You

If you are catching rainwater to water nearby gardens or lawns, you can employ a soaker line, which is basically a hose that is sealed at one end and has holes punctured in it so that water can leak out. You can lay this hose in your garden and allow the water to leak out (only when you turn on the tap, of course). A soaker line is an excellent way to get water to your compost pile, which very often requires water throughout the composting cycle, which we'll discuss in chapter 7.

With larger systems, one of the biggest issues you'll encounter will be water pressure. With smaller systems the pressure will not cause any problems because there won't be enough water in the barrels to cause major pressure buildup. But with a larger system, you will need to consider installing a pump to ensure water pressure in your water lines.

Pressure tanks. Pressure tanks are fairly common and can be purchased at a local hardware store. Normally, these tanks hold about 35 gallons of already filtered water, and they have a bladder inside that fills with water as needed. You usually need a pressure-sensitive pump that will shut off when the bladder has reached a specific pressure. You can use this type of system with a low-flow or solar-powered pump, as it does not require much energy to operate. You do need to ensure that this system is brought inside during the winter because the bladder may be damaged if allowed to freeze.

Pumps. You can use pressure-sensitive pumps as your delivery system as well. They sense the pressure in the water line and kick in when there's a pressure drop. How does that occur? Simple: You open the faucet. That releases pressure and tells the pump to turn on, pushing water through to the tap.

You can also consider inline pump controllers. These are not built into the pumps but are installed outside the system to control the pump when it detects pressure changes. More advanced versions will tell you if they are out of water so the pump does not burn out. This is a good system to use if you want the convenience of constant water pressure.

6. TREATMENT OR PURIFICATION SYSTEMS

Purification or treatment systems are required if you are going to allow any humans or animals to drink the water. Water used for this purpose is called potable. You do not have to treat or purify water that will be used in gardens. Some methods of water treatment are:

- Filtration systems through bags, membranes, or cartridges, all with various levels of filtration, from the basic type that removes parasites to the nanofiltration systems that remove viruses
- Disinfection technologies, such as ultraviolet light and chlorine
- Ozone

See chapter 8 for more in-depth information on water collection and efficient use.

☼ ❄ CLIMATE VARIATIONS

Retrofitting your barn to transform it into a green barn may depend on where you are located and the availability of recycled materials. Choosing supplies that you can find locally will ensure that your barn is not producing extra greenhouse gases through the shipping process.

Subarctic

In the subarctic region, you need to ensure that you are doing all you can to prevent heating leaks by having your ducts tested and sealed, as well as choosing building products that offer the highest insulation rating.

Humid Continental

Whichever green materials you choose should be able to withstand higher levels of humidity without developing mold. Choose materials that are not treated with harmful chemicals and ensure retrofitting takes proper airflow into consideration.

Humid Oceanic

With the higher precipitation rate in the humid oceanic region, you can save yourself a lot of drainage landscaping and ditch digging by catching as much water as you can from your roof rather than having it run off unchecked from the roof into paddocks and pastures.

Highlands

Creating a water-collection system big enough to handle large spring thaws may prove a challenge. Your water-collection cistern temperature will remain above freezing if you build it underground.

Semiarid

You are fairly fortunate to live in an area that requires less "structure" for barns and indoor arenas. You get hot but not too hot, you get cold but not too cold. There are, of course, extreme weather patterns so you don't want to rebuild without taking some into consideration. Have a catchment for freshwater access in the summertime and ensure that your building will protect your horses from the elements in the winter with the correct levels of green insulation.

Arid

Water collection is crucial in arid regions to maximize the amount of water available during droughts. You can use this water to irrigate pasturelands as well as to water your horses.

REDUCE, REUSE, AND RECYCLE

How to have an environmentally sound barn

with minimal consumer waste

WASTE IS EVERYWHERE. Consumers seem to have become too focused on how fast they can get their hands on a product and producers on how cheaply they can make it. Both seem to ignore all the waste produced from excess packaging and the consequences of a "throwaway culture."

THE THREE RS

Reducing, recycling, and reusing are learned habits and behaviors that can become second nature, but we have to practice them regularly. We can change certain habits; for instance, using old clothing scraps as rags rather than paper towels and buying cleaning products or soaps in concentrate so you can mix them in your own recyclable containers.

Set Up a System

The first step in reducing waste on your property is identifying who the waste-producing culprits are. You can control your own waste output, but what about boarders, veterinarians, farriers, and others who come to visit? Set up clearly visible, clearly labeled receptacles for:

- Glass
- Paper
- Plastic
- Food

Don't forget to provide a container for hazardous materials (dewormers, medication, chemicals). Provide heavy-duty sealable plastic bags for each material to avoid mixing potentially toxic substances.

18 Ways to Reduce, Reuse, and Recycle

There are many ways to practice the three Rs on your farm. It helps to ask yourself certain questions before you make purchases or throw something away, such as: "What else can I do with this?" "Will I be able to reuse the packaging?" "Do I really need this?" "What will happen with this product when I'm done with it?"

Here are some simple ways to get started.

1. **Provide ceramic coffee mugs** and old cutlery that you find at the secondhand store, rather than Styrofoam or plastic cups and cutlery, in your viewing area or lounge.

TWO WAYS TO REUSE BALING TWINE

You've seen it at almost every farm you've ever visited. Used to hold together square and round bales, baling (or binder) twine serves a very useful purpose. But it's a one-time-use item and is often accidentally mixed into the hay and may be consumed. Horses generally avoid eating it, but cattle are prone to consuming it, resulting in severe intestinal issues.

CRAFTY TWINE DOORMAT

If you are "crafty," you can collect the twine and reuse it to make doormats:

1. Using a size J hook, crochet a chain long enough for the edge of a standard doormat.
2. Double crochet each row until the mat is as large as you'd like.
3. When finished, double crochet around all four edges, making three double crochets in each corner.

CRITTER BALE-OUT

Braided plastic baling twine with a weight tied to the end can be attached to a fence post and dropped into a water trough to offer trapped wildlife a way out. Often curious mice, squirrels, and gophers climb in looking for a drink and cannot get out. If their bodies are not noticed for longer than a day in the hot sun, they may contaminate the entire tub of water.

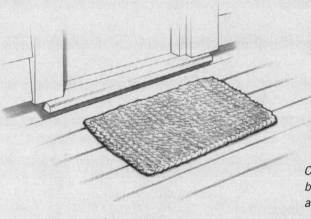

Crochet a sturdy doormat to set before every door in your house and barn.

2. **If you have a toilet** with a 3.5-gallon (13.3 L) flush or greater, place two weighted bottles, bricks, or heavy stones in the toilet tank to displace water. With each flush you will save water equal to the amount dispersed.
3. **Allow bucket baths only,** using minimal soap, on a gravel-filled outdoor spot that will disperse water down a grade and into grassy areas. As mentioned earlier, water used in a washrack becomes "black water"
because of its high fecal content and cannot be reused. Washing your horse outside, however, and letting the water flow into the grass instead of a sewer, means you are using the water a second time by irrigating your lawn.
4. **Avoid environmentally unfriendly drain cleaners:** instead, unclog slow drains in your washrack with a cup of baking soda followed by a cup of vinegar, and let sit.

5. **When planning building projects,** shop at auctions and foreclosure sales for second-hand construction materials.

6. **On hot days** pack your small freezer with half-full water bottles (reusable, of course). When you want a cold drink, top up one of the bottles with water; you won't have to wait for cold water to run from the faucet.

7. **Spot-clean horse blankets** with cold water and a stiff scrub brush instead of sending them for cleaning. Let them dry in the sun.

8. **Install motion-detector lights** on the barn rather than keeping lights on all night.

9. **Clean furnace filters** monthly.

10. **Plant only native grasses and shrubs** that are adapted to your climate, because they will conserve water more efficiently than nonnatives.

11. **Reuse as many containers** as you can that have not held hazardous materials: coffee cans, tin cans, glass jars, barrels . . .

12. **Use grain or bedding bags** as garbage bags.

13. **Collect horseshoes,** clean them, and sell or give them to local artisans who create works of art. If you are talented enough, you can make hooks for hanging bridles, gate latches, or a host of other useful items with horseshoes.

14. **Make an arena harrow** with an old board and a section of chain-link fence.

15. **Rent tractors,** manure spreaders, and other equipment rather than buying them (check with your local conservation district to see if they have a low-cost or free rental program).

16. **Look for used agricultural equipment** (or sell your own) online.

17. **Use scrap wood** to build recycling containers.

18. **Contact wood-pallet companies** to pick up any older pallets you have around your property for repair and reuse.

NEED HELP?

If you live near a large city that doesn't offer recycling collection, recycling companies may be available to pick up your full recycling bins. Coordinate with other local landowners to have recycling bins set up at a common location where everyone can bring their recycling items to be picked up.

Talk to your closest art-and-design college to find artists who use recyclable materials, such as barn wood, scrap metal, burlap sacks, and plastics. They may appreciate many items that you consider garbage.

Seven Tips to Simplify Shopping

Diane Gow McDilda, author of *The Everything Green Living Book* (see appendix B), offers advice about how to cut back on what you buy. Reducing purchases is often one of the hardest new habits to learn because we are conditioned to buy the good/better/best item, even when we really don't need it. Here is an adapted list on how to reduce what you buy for your farm:

1. **Make a list.** Setting priorities as you establish and supply your horse property is important. You can identify your needs and rank them in terms of priorities. If you find yourself craving something new or feel something needs to be repaired, check your list and determine whether you really need it or you need to bump it up on the list of priorities.

2. **Avoid just-in-case purchases.** This is a popular category, especially for older generations who grew up with less. None of us wants to be stuck in a situation where we need something and don't have it. But we aren't talking about an extra hammer or two — do you own a mower attachment for your tractor just in case your ride-on mower breaks down and you are overrun with grass before you can do anything about it?

Bulk Buying

Reducing our consumables takes foresight and planning. We all tend to buy more than we need: sometimes that's a good thing; sometimes it's not. Buying consumables such as grain, feed, soap, and dewormers in bulk can be a great idea, especially if you can buy them with less packaging. For example, using storage bins for bulk grain or feed purchases or large containers for soap that you buy at your local bulk store can reduce your packaging waste considerably. Buying in bulk is always a great option, as long as you aren't overbuying and then throwing out what you don't use.

3. **Consider your space.** Think about where the new purchase will go and what it will displace. Everything you buy takes up space on your farm. And if you are tight for space as it is, new stuff will probably displace something else. In fact, even if you aren't tight for space, do you want to create yet another storage area? To prevent clutter, think about first getting rid of or recycling the item you are replacing. You don't want to find yourself with too many wheelbarrows in various states of disrepair.

4. **Evaluate want versus need.** Take a critical look at what you can do without. Our needs are actually very small, but our wants tend to eclipse them. Ask yourself: Is this a need? Does it improve — or reduce — the safety or health of my environment? Can I do without it?

5. **Beware of bargains.** Tempting though it is, too often a bargain means just taking someone else's junk off that person's hands. A reduced rate on an item that you'll purchase on an ongoing basis, such as bedding or high-quality feed, can be a good opportunity. On the other hand, an extra tractor is not necessarily necessary, especially if the reason it's a bargain is that it's a fixer-upper.

6. **Beware of warehouses.** Again, if you're buying an essential, frequently used item, then by all means check out warehouse deals. But if it's a supersized tub of saddle polish that your grandchildren will still be using years from now, you can probably do without.

7. **Walk to the store.** Or ramble around your farm. Use your feet while you can, and save the scooters and gas-powered vehicles for later in life. It's not the worst thing to cut across your own pastures — you probably need to visit them more frequently anyway.

HARMFUL PLASTICS

Over the past few years, attention has focused on plastic containers and materials and our dependence on them for everything from food and beverage containers to plumbing and building materials. Plastic is a nonbiodegradable product that stays in landfills a long time, all the while leaching chemicals into the soil and eventually into our water supplies. Most plastics are manufactured from petroleum, a nonrenewable resource.

What's the Problem? Evaluating Plastics

Plastics are assigned numbers based on their chemical constituents. Numbers 1 through 7 each pose possible health concerns for humans — and likely for horses. Here's what the numbers identify. Resin identification codes are developed by SPI: The Plastics Industry Trade Association (see Resources) to indicate use, hazards, "recyclability," and consequences of use.

The number embossed on a plastic container indicates its chemical makeup and recyclability.

PETE #1 (POLYETHYLENE TEREPHTHALATE)

Uses. Soft drink, juice, water, detergent, cleaner, and peanut butter containers; not designed for reuse, but they often are reused (especially water bottles that you refill, not realizing that they are a "single-use" product)

Hazards. Break down easily, leaching chemicals into your body; reusing these containers causes microscopic scratches and cracks, which serve as reservoirs for bacteria

Consequences. Contains phthalates, which increase the risk of reproductive cancers and infertility

HDPE #2 (HIGH-DENSITY POLYETHYLENE)

Uses. Opaque plastic milk and water jugs, bleach, detergent and shampoo bottles, and some plastic bags; generally considered safer for storing food and water

Hazards. Not applicable

Consequences. Should be recycled, but many are not and therefore clog up landfills and waterways (specifically the plastic bags)

PVC #3 (POLYVINYL CHLORIDE)

Uses. Made of vinyl and used in cling wrap, some plastic squeeze bottles, cooking oil and peanut butter jars, detergent and window cleaner bottles

Hazards. No food or drink should ever be stored in these containers

Consequences. Contain dioxins, which contribute to lung cancer and endocrine and autoimmune conditions, and phthalates, which cause liver, kidney, and testicular damage

LDPE #4 (LOW-DENSITY POLYETHYLENE)

Uses. Grocery-store bags, most plastic wraps, and some bottles; generally considered a safer plastic

Hazards. Choking hazards for sea life if not disposed of properly

Consequences. Should be recycled, but many are not and therefore clog up landfills and waterways (especially the plastic bags)

PP #5 (POLYPROPYLENE)

Uses. Most Rubbermaid, deli soup, syrup, and yogurt containers, straws and other clouded plastic containers, including baby bottles; considered a safer type of plastic

Hazards. Not applicable

Consequences. Not applicable

PS #6 (POLYSTYRENE)

Uses. Styrofoam food trays, egg cartons, disposable cups and bowls, carryout containers, and opaque plastic cutlery

Hazards. Especially toxic when heated (e.g., when used for coffee or takeout)

Consequences. Contains toxins that cause reproductive problems and cancer

OTHER #7 (POLYCARBONATE)

Uses. Most plastic baby bottles, 5-gallon (19 L) water bottles, "sport" water bottles, metal food-can liners, clear plastic "sippy" cups, and some clear plastic cutlery; new bio-based plastics labeled #7 are safe to use

Hazards. Contains bisphenol-A, which disrupts hormones and mimics estrogen

Consequences. Minute amounts of this substance can result in reproductive disorders such as infertility, endometriosis, fibroids, low sperm count; prostate, breast, uterine, ovarian cancer; hypo- or hyperthyroidism; early puberty; hyperactivity; obesity

---

Controversial Bisphenol-A

In 2008 Health Canada released information on its study of bisphenol-A:

> The draft screening assessment proposes bisphenol A is "toxic" to human health and the environment, as defined in the Canadian Environmental Protection Act, 1999. This preliminary assessment tells us the general public need not be concerned. Our focus is now on newborns and infants (under 18 months). Science tells us that exposure levels are below those that could cause health effects, but since they are close to the levels where potential effects could occur, the Government wants to be prudent and reduce exposures further. With respect to ecological effects, our initial assessment shows that at low levels, bisphenol A can harm fish and organisms over time. Studies also indicate that it can currently be found in municipal wastewater.

As of January 2009, 14 states are also considering a ban on BPA. A study conducted by Richard Stahlhut at the University of Rochester was released in early 2009 and found that the chemical stays in the body much longer than previously thought. The National Toxicology Program in the United States has expressed some concern about the safety of the chemical for infants and children, but the Food and Drug Administration maintains that it poses no risk; however, the FDA's science advisory board has since recommended that the ruling be reconsidered.

Plastics and Your Horse: Making Smart Choices

While humans should avoid many of these plastics, what does it mean for your horse and your barn? Plastics labeled 3, 6, and 7 are often used around the barn for storing or feeding food (scoops for grain, cups for coffee) but can be easily replaced by ceramic or glass containers. Or use the new bio-based plastics, such as polylactic acid, or PLA, a corn-based plastic-and-foam laminate made from potatoes, corn, rice, or tapioca.

PVC: The Toxic Plastic

The most common plastic found in the barn is the popular PVC pipe, sometimes used to help guide water flow throughout the property in underground culverts or in conjunction with watering systems. It was developed for widespread use during the Second World War for use on U.S. military ships.

According to the Institute for Agriculture and Trade Policy (see Resources), however, PVC is now known as one of the most harmful plastic products that we use.

Polyvinyl chloride, also known as vinyl or PVC, poses risks to both the environment and human health. PVC is also the least recyclable plastic. Here are some facts about it, from reports of the U.S. Environmental Protection Agency (see Resources):

- **CANCER RISK**. Vinyl chloride workers face elevated risk of liver cancer.
- **POLLUTION**. Vinyl chloride manufacturing creates air and water pollution near the factories.
- **TOXIC ADDITIVES**. PVC requires additives and stabilizers to make it useable; for example, lead is often added for strength, while plasticizers are added for flexibility. These toxic additives increase pollution and human exposure.

- **DIOXIN.** Dioxin in air emissions from PVC manufacturing and disposal or from incineration of PVC products settles on grasslands and accumulates in meat and dairy products and ultimately in human tissue. Dioxin is a known carcinogen, and low-level exposures are associated with decreased birth weight, learning and behavioral problems in children, suppressed immune function, and disruption of hormones in the body.

ALTERNATIVES

Take heart: there are alternatives to PVC piping. Concrete is an ancient alternative, used by the Romans to create their famous aqueducts. It can be poured to create conduits for water (there is a higher environmental cost in the creation of concrete), as well as storage tanks and cisterns.

The most popular alternative is high-density polyethylene (HDPE), which is nonchlorinated, requires fewer additives, and has a much higher recycling rate. It can be used in virtually the same manner as PVC pipe but has far fewer toxic consequences.

LIGHTEN UP!

Conserving energy from light and electricity use has been widely known ever since the first father shouted "turn off that light, we're not made of money!" Now, thankfully, we can conserve energy even when the lights are still on.

Compact Fluorescent Light Bulbs

New legislation introduced in the United States means that by 2012 incandescent lightbulbs will no longer be available. Many consumers have protested because compact fluorescents (CFLs) are currently quite a bit more expensive. Their life span is significantly longer than that of incandescents, however, so over the long term you'll save.

When CFL bulbs were first introduced consumers weren't sure what to make of the "spirally bulbs," but they soon warmed up to them when they started seeing the cost savings.

What to Do If Your CFL Breaks

1. First, leave the room to ensure you don't inhale any vapors. Open windows, and keep everyone away from the area. If you are in the barn, you may want to take the horses outside for 15 minutes.
2. Wear rubber gloves and scoop the powder into a plastic bag (preferably a sealable bag) with a stiff piece of paper or cardboard. Don't use a broom, as it will send dust up into the air.
3. Use a damp cloth or paper towel to wipe everything down. Dispose of the cloth by sealing it into the plastic bag. Place the sealed bag inside a second plastic bag and seal that. Set the bag aside to take to the depot that accepts CFLs.
4. Wash your hands thoroughly.

THAT ANCIENT BARN FRIDGE

Raise your hand if you have an old "barn fridge" in a tack room, a dusty, ancient appliance that came out of Grandma and Grandpa's house when they moved and has lived in the barn longer than many of your animals. That fridge needs to go. Old, inefficient appliances like this consume energy desperately, as if the power were going to be turned off tomorrow.

Your best choice is a compact fridge with a high Energy Star rating. Keep your fridge well stocked to help conserve energy. Add bottles of water and ice or gel packs, which are handy in case of emergency. They take up room in the fridge and freezer, allowing the motor to work more efficiently to keep the temperature cool.

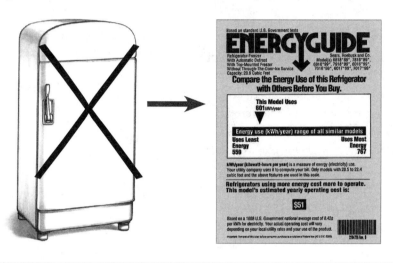

As they come into greater use, we will have to get used to disposing of CFL bulbs properly. We routinely chuck incandescent bulbs into the garbage bags destined for landfills, but CFLs contain a small amount of mercury gas — not enough to pose health risks in your home but enough that if we all threw our bulbs in the garbage, we'd have a serious issue with mercury contaminating our landfills.

You can locate a CFL recycling depot online (see Resources), or find a disposal site in your area. Some retailers — IKEA, for example — allow you to return the CFLs to the store for safe disposal. Just make sure you are transporting the remains of the bulb in an airtight container.

Lighting Indoors and Out

Using light efficiently often has more to do with barn construction than it does with the actual lighting used. We discussed this topic in chapters 4 and 5. Designing a barn to maximize the light from the sun will help reduce your electricity bill because you won't have to turn on so many lights.

You may need to light the outdoors: perhaps you ride in the evening or walk horses back to their paddocks when the sun has already gone down. Keep outdoor arena lights on sensors that turn on only when required but can be turned off and unplugged when not needed.

For lighting walkways you can install small solar lights. Available at any gardening store, these fixtures are about a foot (30 cm) tall and are stuck into the ground on plastic spikes. They have small solar panels on top that store solar energy until they sense that it is dark enough to light up.

Another option for efficient lighting is light-emitting diode (LED) lights. These produce more illumination per watt than incandescent bulbs and therefore work very well with low-power solar- or wind-generated systems.

When you need to make a trek to and from an outdoor building, you could also consider using a flashlight. You can find LED crank flashlights at most camping and outdoor stores. In Canada they are available at Canadian Tire and in the United States you can find them in many retail and online stores.

MONITORING WATER CONSUMPTION

Obviously, your water consumption will depend on the number of horses you have in your barn, the temperature, and how long they stay inside.

To ensure that you are not wasting water, regularly maintain all automatic waterers. These have long been considered the most efficient watering method, since only a small bowl is filled as the horse drinks, and clumsy horses are not spilling entire buckets of water in their stalls. But if your waterer is not properly maintained and drips steadily, you may be wasting more than a few spilled buckets of water.

Very often, automatic waterers that drip and leak are simply in need of a thorough cleaning. Horses that try to dunk their hay in their waterer leave behind bits of hay and feed, clogging up the floats that determine when the water requires turning on or off. A clogged or weighted-down float may cause a continuous trickle of water.

If you use tank waterers, make sure that you have covered about 75 percent of the top

Unplug from the Grid

You may not realize it, but almost every appliance that you plug into an outlet draws energy even when turned off. This is called a "phantom load." Bulbs, water heaters, radios, fans, and other items need to be unplugged when not in use. Radios in barns are frequently left on overnight by well-meaning horse owners who believe their horses find the sound soothing. Actually, it's anything but. Horses rely on their hearing in the same way that we rely on our sight. Radio noise in a barn, to a horse, is what trying to sleep in a room with a strobe light is to a human.

of the tank, allowing the horses to drink from the remaining 25 percent. This ensures efficient use of the water in the tank and reduces evaporation.

Also make sure that all water fixtures, from your hose connections to bathroom faucets to washrack taps, are in good repair.

Ecofriendly Toilets

Busier boarding barns often need to provide bathroom facilities for their customers. If you have an indoor washroom in your barn, you may want to consider switching to a water-saving toilet, such as a dual-flush model. Studies have shown that these toilets, which offer you a choice of a short 0.8-gallon (3 L) flush or a long 1.6-gallon (6 L) flush, can reduce flush volumes by more than 50 percent in public facilities and almost 70 percent in residential homes.

Tankless Water Heaters

A great choice for larger barns — particularly show barns — that require washrack facilities is the use of a tankless water heater. A regular-sized

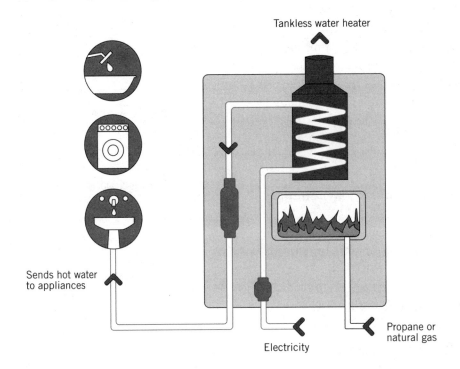

Tankless water heater

Sends hot water
to appliances

Electricity

Propane or
natural gas

*The tankless water heater operates very quickly and heats water
only when you need it, rather than keeping it hot all day long.*

water heater for use in home or barn can take up 90 square feet (8 sq m) of space and waste a lot of energy. Water heaters fill up with water; hold the water at a specific temperature until it is used; and refill, reheat, and store at that same temperature. Many barns can survive without heated water, but the farther north you live, the more likely you are to have a need for heated water for bathing, washing, and cleaning.

A tankless water heater runs on propane or natural gas (for larger quantities of water) or electricity (for smaller quantities) and flash-heats the water on demand, rather than maintaining the hot water at a specific temperature for hours on end. This makes good sense. Would you let your car run all night so you can drive it to work the next morning? Then why would you heat your water all week just so you can use it once or twice?

The tankless water heater uses a flame to heat copper tubing, maximizing the surface area that comes into contact with the flame, thus allowing

for the greatest heat-transfer efficiency. The water is heated very quickly and effectively.

Even when using a tankless water heater, you still want to ensure that you are using the warm water most effectively by keeping the heat set at the lowest setting you require. It's much more efficient to deal with cooler water than to start with water that is too hot where you have to cool it down by added cold water to it. Why not just start out with a nice warm temperature?

☼ ❄ CLIMATE VARIATIONS

There are really no major climate considerations for reducing, reusing, and recycling in North America, as the processes are generally universal. During cold seasons your biggest energy-reduction impact will be reducing your reliance on fossil fuels to heat your barn. Building a barn to minimize heat loss is an excellent start. See chapters 4 and 5 for more information on efficient barns.

III

Eco-Equine
Management

MANAGING A BARN and everything that goes into it is no small undertaking. You have a herd of horses that rely on you for their food, shelter, and water, and sometimes you even have customers that expect a certain level of comfort when they come to ride. Establishing your facility as a "green barn" goes beyond the structures that we have covered previously. Now let's start talking about the everyday practices that help change the destructive habits that our throwaway society has developed.

MANURE MANAGEMENT

Emphasize reduction and reuse

MANAGING MANURE is the number-one challenge for horse properties of any size. The more horses you have, the larger the challenge. A single horse can produce 50 pounds (23 kg) of manure a day. That's 9 tons (8 metric tons) of manure a year!

Your county or local regulating body probably has established rules for the number of horses you can have on your property, and one reason for this is the amount of manure horses produce. Whether your horses live in stalls and manure is removed each day or they are kept outside and the manure stays where it falls, you can calculate the amount of manure your facility produces each year. Just multiply the number of horses by 50 (or 22.7 in Canada) to find the pounds (or kilos) they produce per day.

COMPOSTING MANURE

For many years, equine facilities around North America have spread their manure on fields surrounding their property. The reasoning was that manure is fertilizer, and fertilizer is good for growing things. Unfortunately, many well-intentioned horse people have been fertilizing the wrong way.

Composted manure is great for fertilizing everything from grass and hay to mushrooms and herbs. Manure straight from the stall, full of bedding and possibly parasites or chemicals, is not. When left where it falls out in the pasture, manure will eventually decompose, but the practice of composting hits the fast-forward button. As a bonus, the composting process does away with parasite eggs and larvae. It also ensures that weeds are not spread throughout your pasture just because your horse ate a few seeds, didn't digest them, and has now planted them around your property through his manure.

Not only is composting healthier for the land, but it can also take a daily horse by-product and turn it into a money-saving and money-making product. When it's done correctly, you will replace a big, smelly manure heap with a fertile pile of dark, earthy-smelling compost, which any gardener would love to use, at half the volume of uncomposted manure.

Just what are the benefits of compost? When manure is properly composted and added to soil it:

- Provides macronutrients such as nitrogen, phosphorus, and potassium in forms that are easily used by plants
- Provides micronutrients such as manganese, iron, zinc, copper, and boron (which are not usually found in commercial fertilizers)

- Uses a slow-release method of fertilization that isn't likely to result in overfertilization
- Has high levels of organic matter, which increases the soil's ability to hold water and retain nutrients
- Lightens heavier soils (such as clay) to allow more air and water to access root systems
- Results in soil that is more stable and less affected by erosion
- Stretches your budget by reducing your fertilization costs

HOW MANY HORSES CAN I HAVE?

Many counties in both Canada and the United States restrict the number of horses you can own — normally to about "one animal unit" per acre (0.4 hectare). Often, you can apply to house more if you have indoor stabling for the night and outdoor paddocks or pastures for the day. A mare and foal normally count as "one animal unit" until the foal is weaned.

Ten horses on one acre can consume as much grass in an hour as one horse on one acre in a week.

Composting Basics

When manure is composted, it relies on microorganisms such as bacteria, fungi, insects, and worms to break it down into smaller particles. The process recycles nutrients into the soil while releasing carbon dioxide, water, and heat. These helpful microorganisms need air, water, and food to survive.

Air. You can easily allow air into your compost pile by leaving some bulkier materials, such as bedding, in the manure. You can add lawn clippings, leaves, or old hay. Air can also get where it's needed when you turn your compost pile regularly or aerate the pile by inserting a PVC pipe with holes punched in it.

Water. The perfect moisture level for our friends the microorganisms is about 50 percent. Pure manure has a water content closer to 70 percent, but adding bedding and other bulkier materials reduces that level.

Temperature. When microorganisms work, they create heat. The warmer the pile gets, the faster it will decompose. The most efficient temperature is between 130 and 150°F (55–65°C). The bigger your pile grows, the more heat it will retain. Smaller piles are colder.

Food. The microbes require carbon, nitrogen, and other nutrients from compost additives (hay, bedding, straw) so they have energy to work. Nitrogen is easily obtained from the manure, while carbon is found readily in bedding.

THE ALL-IMPORTANT C:N RATIO

The key to successful composting, however, is the ratio at which these elements occur. The carbon-to-nitrogen (C:N) ratio should be 30:1.

C:N ratios do not refer to the actual weights. There is 1 pound (0.5 kg) of carbon in 20 pounds (9 kg) of horse manure, but 500 pounds (227 kg) of sawdust produces only 1 pound of nitrogen. Undiluted horse manure contains modest levels of nitrogen (1 to 2 percent of dry weight) and supplies small amounts of nitrogen to plants via

---

Manure Management Timeline

Your manure-pile management timeline will vary according to your needs: how many horses you have; how much time you have to invest in its upkeep and your field cleanup; and whether you plan to market compost as a by-product of your facility, as there are specific times of the year (most notably, spring) when the demand is highest.

DAILY

- Pick up stalls
- Pick up manure as it falls in arenas and riding rings to keep your sand rings less dusty
- Sweep or rake alleyways and barn aisles

WEEKLY

- Empty manure buckets in arenas and barns
- Pick up manure in turnout paddocks

MONTHLY

- Clean out paddocks; do one or two a month, and rotate through all paddocks
- Pick up any manure piles near waterers or feeding areas in fields and pastures

YEARLY

- Harrow pastures and fields once or twice a year to break up manure
- Clean up large areas where manure collects

---

slow release. The addition of bedding, however, reduces the amount of nitrogen available to plants through something called "microbial immobilization," which basically reduces the uptake of nitrogen to any plants. To minimize this, try to maintain one part manure to two parts of bedding, by volume, to get the right ratio for efficient composting.

If you add too much bedding to the manure pile, it tips the ratio too heavily in favor of carbon and means you need to supplement with nitrogen to have effective compost. This is especially true if you use wood chips or sawdust, because they are so carbon heavy. Wood chips will provide extra bulk and allow for better airflow, but if you manage sawdust correctly, it will actually compost at a faster rate. You'll know you need to add nitrogen if your compost pile is not heating up to the appropriate temperatures (see page 103).

Choosing Your Site

The location of your compost pile will also be essential to its success. Take these points into consideration:

Rules and regulations. Check with your local regulating body for rules about how close your compost pile can be to springs, wells, or open bodies of water. (See box.) The larger the body of water, the greater the likelihood that there are rules governing what can be near it.

Flat and dry. Find a flat, somewhat elevated area that is easy to access. Avoid low points on your property: pooling water will attract flies, and it will stink. While you don't want water to pool, neither do you want runoff.

And when you find a nice, flat area for your compost pile, make sure it's not a nice, flat flood plain. One flood can wash away an entire manure pile and contaminate everything the water touches.

SITING A MANURE PILE

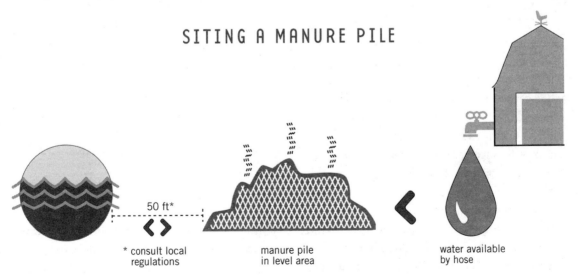

50 ft*

* consult local regulations

manure pile in level area

water available by hose

The ideal location for your manure pile is on flat ground, set back from any water but with water available, and close enough to the barn to be convenient.

Water access. In case you need to add water to reach the correct moisture level, choose an area where you can access water for a sprinkler or hose. (Your water catchment will work wonderfully for this.)

Convenience. To manage the compost pile efficiently, make sure you leave enough room to maneuver equipment around it for bringing manure to the pile, removing the compost, and turning the compost when required.

METHODS OF COMPOSTING

In the following composting methods, you'll see that the methods that require a little more maintenance are the ones that generally turn out a higher-quality product.

Freestanding Compost Piles

This method is the easiest but not necessarily the best. It simply involves piling the manure in one area and turning it over three or four times a year. Because it's a simplified composting method, it may take upward of a year to turn the entire pile to compost. In addition, it may not adequately kill the parasites and weed seeds. However, it's a passable choice for those with just one or two horses on their property.

Windrow Composting

This method is very efficient for those using manure spreaders because it utilizes space in a pasture or field and allows for easy tractor maneuvering. Wheelbarrows work fine, too, if you don't mind a little walking. You form a windrow by piling the manure in a straight line, either by driving the manure spreader along the same path or piling it in a row with your wheelbarrow. An ideal windrow is 5 to 6 feet tall (1.5–1.8 m) and 6 to 10 feet wide (1.8–3 m), so you will spend a good part of the year driving over the same spot and gradually building up the pile or dumping the wheelbarrow in a row and pushing it up higher with a tractor bucket.

The method of driving over one spot for months is effective, but after you've been doing it for a while it can be daunting to drive a tractor up a large, soft hill. You need to be very careful that you do not tip over. Instead, consider spending two or three months driving on one path, move to the side, and make a second row. Then push the two together.

The biggest downside to this method is that you will see horses meandering through the long rows, sometimes pawing and picking out bits of leftover hay, like dumpster divers behind a restaurant. This is not an ideal situation because

Freestanding manure piles are easy to maintain but slow to become compost.

Windrow composting is efficient if you have more than six horses and plenty of extra space. The pile must be turned several times a year with a tractor or front-end loader and does not heat up as quickly as with other methods.

along with those bits of hay, they may be picking up parasites and worms from other horses. You might consider sectioning off the windrow with electric fencing. (The buffet is closed!)

The windrow can be turned with a tractor or front-end loader several times a year and is very efficient for larger herds of horses if your facility has the space to compost this way. You need to monitor the compost in the same way that you do the other methods, checking for moisture and temperature levels regularly. Windrows need to be larger because they need to build up more heat. Being left out in the open field and subjected to the elements means that they will not heat up as quickly and will therefore have a longer "cooking" time, possibly over a year, depending on the size.

Bin Composting

One of the best methods in terms of efficiency, environmental impact, and labor is multiple-bin composting. This involves having sheds or shed rows with at least two separate compartments, possibly more. Each bin is a three-sided enclosure that allows access to the bin from the fourth, open side.

In the first bin you will pile your manure until it is filled to capacity. About once a month you will turn this pile either by hand or by machine. When the first bin is filled, you begin to fill the second, then the third. If you have a large facility, you might consider four or more bins. With this method the smaller, enclosed piles can compost at a faster rate so that, ideally, the first bin is ready to be emptied for use by the time you have filled the last bin.

To make sure that you are adequately turning and airing the compost, move it from one full bin to another and allow it to compost in the second bin while you return to filling the first bin again. When that bin is filled, shift the pile down to the next bin, which aerates it once again. Turning so that all surface material is aerated is crucial to the process. The better you

CALCULATING BIN SIZE

First, determine the amount of manure you will need to hold, and then decide how long you're going to keep the manure there. If you have only one or two bins, you must store the compost longer than if you have four or five bins to use. If you do have just one or two bins, you might consider a six-month storage time; if you have four or five bins, you could go with three months at a time. Keep in mind that the volume of the compost decreases over time.

In the following equations,

A = How many horses you have

B = Average output for an 1,100-pound (500 kg) horse is 0.8 cubic feet (0.02 cu m) a day

C = Length of time compost will remain in bin, in days

D = Volume (cubed) of bin required

$$A \times B = AB$$
$$AB \times C = D$$

When you reach your answer, D, it will be in cubic feet (cu m). This will help you determine the size of bin you require in height, width, and length (H × W × L). You may need to guesstimate a little, especially if you will include soiled bedding or if you regularly strip your stalls, which could double the space requirements. For example:

A × B = AB: 10 horses × 0.8 cubic feet = 8 cubic feet

C: Manure will accumulate for 3 months (90 days), so 8 cubic feet × 90 = 720 cubic feet at maximum capacity.

D: To compensate for bedding and stripping of stalls, let's say that your bin should be 1,200 cubic feet (34 cu m). This is 10 feet (3 m) tall × 10 feet wide × 12 feet (3.7 m) deep. For practical reasons two bins, each 5 feet (1.5 m) tall × 5 feet wide × 6 feet (1.8 m) deep, would provide the capacity you'll need.

"stir" the pile, the more consistent the compost will be at the end.

The size of your compost bins will depend on the amount your horses produce, the length of time you can store the compost, and what type of equipment you'll use to turn and move the manure. In an ideal situation, you don't need to clear a bin and haul away the finished compost for several months. This allows enough time for complete composting.

CONSTRUCTING THE BINS

How you build your bins will be determined not just by how much manure your horses produce but also by what type of equipment you use with your compost.

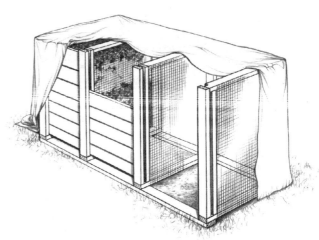

Bin composting makes sense if you are short on space and have only a few horses.

Smaller bins that are turned by hand can have either a dirt floor or a slatted wood floor built with untreated 2×4s. The wooden floor will allow for better airflow to the pile. A dirt floor will work as well.

If you are using mechanical means to turn your compost pile, however, a dirt or wood floor may be a disadvantage. It is too easy for a front-end loader or tractor to accidentally dig a hole in the ground over time. Slatted wood floors are better than dirt as long as you take care not to damage the wood when turning and moving the pile. The best choice is a flat cement pad that will allow you to scrape up all the manure when you are moving it.

The walls will also need to be sturdy, with heavy-duty posts and 2×6 boards on both sides, to withstand the weight and pressure of the manure against the walls.

Cover the bins with a roof or plastic tarps. This helps protect your compost from rainwater and snow so you can properly regulate the moisture content. If the piles became too saturated with snow or rain, the minerals you are doing such a great job to create and protect will be washed away and may cause a hazard if they leach into the groundwater. This protection from weather is the main reason structured composting (in bins or storage containers) is a superior method to windrow and freestanding methods.

The path leading up to the bin needs some sort of gravel to prevent it from becoming a mud pit, especially if you use mechanical equipment to move the compost. You can quickly damage this area if you drive over it too soon after a rainfall or snowmelt.

Compost Must-Haves

A way to turn the pile. If the pile is small enough, a shovel and your own labor (or that of your teenage kids) will do. A tractor or front-end loader may be required to completely turn larger piles.

Water supply. Water is best provided with a hose and sprayer, but a portable water tank with a pump on the back of a truck will also work.

Adequate space for storage and composting. An enclosed area or a pile big enough that it can maintain its own internal temperature is essential. Smaller individual piles cannot maintain high enough internal temperatures for true composting.

Compost Caretaking

Now that your manure is becoming compost, you need to do some maintenance to make sure it turns into a high-quality product that you can use on your own farm or sell to others (more on that later). If you ignore your compost pile, you're putting a lot of effort into something that is not going to be worth a thing.

To ensure your compost is reaching the perfect temperature, it should be at least 35 cubic feet (1 cu m). You can add to the pile as you go along, especially if you use the bin system. You can also add manure and bedding when you clean out your horse trailer or pick up manure from riding arenas or fields. Keep dirt and trash out of

---

Compost Precautions

It is unlikely that you will catch any diseases from your horse by handling compost or manure. However, some bacteria and protozoans (such as *E. coli* and *Giardia*) can pass from horse to human or can be transmitted to water sources if you are not taking care to prevent runoff.

Parasites are rarely a problem unless the compost has not been kept at a high enough temperature to kill both parasites and their eggs. Only straight manure and inadequately prepared compost pose a risk.

---

the pile, because they will degrade the quality of the compost. That plastic bag or length of baling twine is not going to disappear in the compost pile; it's just going to stay trash.

TAKING YOUR PILE'S TEMPERATURE

As mentioned before, the perfect composting temperature is between 130 and 150°F (55–65°C). It can take some time to reach that heat, depending on the weather and the size of the pile. In winter it may take a month to reach an appropriate temperature. Turn the pile much less frequently or not at all during the long stretches of deep-freeze temperatures, like those on the northern prairies.

This temperature level should be maintained for at least three weeks to make sure that efficient breakdown is occurring and that parasites and weeds are killed. On the other hand, if the pile gets too hot, the microorganisms that are trying to do their work are also killed.

Taking the temperature of a manure pile is quite simple. Tape a thermometer to the end of an old pitchfork or shavings fork handle and push it into the middle of the pile. A digital thermometer works wonderfully, as there is little chance of breakage. (Compost thermometers can also be found at gardening, hardware, or farm-supply stores.) Leave it in for a minute

A thermometer will help you monitor the heat of the pile and know whether the composting process is working.

(most digital thermometers beep when they are done, but you won't hear it from inside the pile, so know the duration of time before the beep), and then remove it. Do this a couple of times a week. If it's getting too hot, simply turn the pile over, and allow air to cool it.

TURN, TURN, TURN

Turning the pile to aerate it regularly is a crucial component of the compost process. How often you need to turn a pile depends on the weather outside and the heat of the pile. If the internal temperature is below 110°F (43°C) or above 150°F (65°C), it needs to be turned.

If you are using a bin system, simply turn it over while you're moving the pile to another bin. If you find that you aren't able to turn it as often as you'd like because the bin isn't filling up quickly enough, add a perforated PVC pipe as mentioned earlier. This will allow cooler air to get into the pile.

TESTING MOISTURE

The moisture level of the pile is also important. If it's either too dry or too wet, the microorganisms will be unable to work. Testing the wetness is as simple as reaching in and grabbing a handful. The compost should feel soft and damp, but when you squeeze it, no water should be expressed.

Is It Done Yet?

Average composting times vary by region. In colder climates it may take two to three months in the summer and four to six months in the winter. In warmer climates two to three months is the average no matter what the season.

Here's how to tell if your compost has finished "cooking":

- It no longer heats up after being turned
- It has an earthy smell
- It has a crumbly texture
- You see no evidence of weeds or parasites

🍃 COMPOST WISDOM 🍃

If your compost pile is too wet, add some old hay, leaves, bark, or other dry biodegradable material. These will act like a sponge, preserving and slowly releasing the excess water.

If your compost pile is too dry, sprinkle it with water and turn it. But be careful, since adding water to a compost pile is like adding salt to soup: too much, and you wreck the whole thing. Wet it down as you turn it each time, and monitor the moisture level. If it's consistently dry because of weather conditions, wet down the manure in your wheelbarrow before you add it to the pile.

TIPS AND TIMESAVERS

To add nitrogen you may need to supplement with a little bit of nitrogen-rich fertilizer, such as urea or ammonium nitrate. Don't overdo it; if your horses produce about 50 pounds (23 kg) of manure a day each, you need only 2.5 ounces (70 g) of urea or 3.25 ounces (90 g) of ammonium nitrate per horse each day. This can be mixed in once a week.

You can also add coffee grounds to your manure pile as an additional source of nitrogen (about 2 percent nitrogen by volume and in a 20:1 C:N ratio). Remember, though, that coffee grounds take at least two weeks to compost and uncomposted coffee grounds are not good for plants — it stunts their growth. (Now, the studies were done on lettuce, but it would seem that the same effect would apply to other plants.)

To help speed up the composting process, consider mulching your manure first with a shredder or tub grinder. If you reduce the particles before adding manure to the pile, more surface area allows the microbes to get to work faster.

BEWARE OF FIRE!

Although rare, it is possible for your compost to catch fire. This can occur if the center of the compost gets very hot while the outside stays dry. This is the number-one reason that you should not ignore your compost pile.

VERMICOMPOSTING

Vermi is Latin for "worm," and vermicomposting is a process that puts worms to work. The digestive tracts of certain species of earthworms (e.g., red worms, tiger worms, red wigglers) can work with compost organisms to decompose manure and bedding.

Pros and Cons

This method is slightly more expensive than the regular processes. Nevertheless, it often results in a more fertile compost that takes less manpower (and mechanical power) to complete, without the cost of extra nitrogen.

Vermicomposting is ideal for manure piled in windrows. Because the worms basically turn over the compost themselves, minimal tilling and turnover are required.

There are some downsides, however. The initial cost of red worms can be up to $12 per pound (0.45 kg) for bulk orders. You may need 100 pounds (45 kg) of worms per horse, though some species double their population every four months and can be collected and reused

The most popular type of worm for vermicomposting is the epigeic (Greek for "upon the earth") type. He (well, he/she, since they are hermaphroditic) lives on the surface of the soil and does not form any permanent burrows (why settle down when you are your own partner?) while feeding on decaying organic matter. This type of worm is variously called red worm, manure worm, brandling worm, red wiggler, and compost worm (all *Eisenia fetida*); tiger worm and red tiger worm (both *Eisenia andrei*). But all are referred to by their generic name *Eisenia* (eye-sen-ee-uh). CSU details specific conditions that these worms prefer:

- **TEMPERATURE.** Tolerate 39 to 90°F (3.9–32°C). Prefer 65 to 75°F (18.3–23.9°C).
- **MOISTURE.** Tolerate moisture levels from 40 to 100 percent. Prefer 60 to 80 percent.
- **ACIDITY.** Tolerate a pH range of 2 to 9. Prefer a range of 5.5 to 7.
- **LIGHT.** Sensitive to both sun and electric light. Breathing slows and becomes shallower, and they are disoriented. Just 30 minutes in the sun can kill them.
- **DENSITY.** Tend to live in concentrations of fewer than 1,000 worms per cubic foot (0.3 cu m) of material. Will seek ideal conditions in a windrow if there is food.

for another compost pile. You need to monitor and care for your growing family of worms year-round, and because the compost pile does not reach the high temperatures of regular compost techniques, the technique may not kill weed seeds or parasites.

It's All in the Castings

According to Colorado State University (CSU), the key to vermicomposting is the "castings":

Worm excrement is commonly called castings. While they may look and feel like tiny flecks of sticky soil, they are full of beneficial soil microbes. Scientists have yet to conclude exactly why *Eisenia* [genus name for certain worms] castings are good for plants, but they seem to contain nutrients that plants can easily use and disease-suppressing microbes. The mucous covering on the castings also appears to slow down nutrient release. In addition, enzymes in the gut of *Eisenia* may kill many pathogens harmful to plants, horses, or humans that pass through its gut. In any case, castings will not burn your plants, even seedlings, and they have a neutral pH.

So where to begin? First, start with as many worms as you can afford, and maintain them as they grow. Keep track of the time it takes your "worm herd" to double by taking random samples of the dirt, counting the worms and estimating the number of worms per pound (or kg) based on the concentration in your sample. Keep an eye out for the development of castings, which are darker than manure and signal that the worms are eating and healthy. While you don't want to dig in and disturb the worms on a daily basis, a weekly check is a good idea. Even taking a quick digital picture for reference can help you track their development.

Starting a Vermicompost Windrow

Colorado State University offers two methods for establishing your vermicompost windrow in a fact sheet titled "Composting Horse Manure in Dynamic Windrows" (see appendix B).

OPTION 1. Establish the area for the vermicomposting windrow, but do not add additional nitrogen to the manure and bedding. Monitor temperature in the windrow, and turn it before temperatures reach 145°F (62.7°C) to prevent carbonizing the material and reducing its palatability for *Eisenia*. The worms will digest your manure faster if the material has composted for 7 to 10 days first.

Create an initial base of material 6 feet (1.8 m) wide, 18 inches (46 cm) tall, and 6 feet long. The base should be oriented east-west so it receives sunlight on its south side all day. Water this base of material until it feels wetter than a wrung-out sponge. Divide your quantity of worms, and distribute them evenly over the top of the moistened base material. *Eisenia* should immediately migrate into the material.

Once the worms have colonized this material, add a 3-inch (7.6 cm) layer of material weekly to the start-up pile, and moisten appropriately. Add to the pile gradually to prevent it from heating up and killing the worms. (They prefer temperatures below 90°F [32°C].)

Since this process should not generate heat, it is advisable to set up the initial windrow in the summer. A large windrow will be formed

during the season to help retain heat during the winter months.

After the pile reaches a height of 3 feet (0.9 m), add new material to the end of the pile in the direction that the windrow will be built. It should match the width of the start-up pile and have a height of 3 feet to allow for volume reduction. Facilitate the heating of this new material (by turning and watering) to a maximum temperature of 145°F (63°C). *Eisenia* will migrate into this new material once they have digested all of the start-up pile and the new material has cooled below 90°F (32°C) and has ideal moisture levels.

Continue lengthening the windrow until you run out of space. At this point you can make a

HOW MANY WORMS DOES IT TAKE?

Optimal amount of *Eisenia* required to digest all manure and bedding as it is produced.

NUMBER OF HORSES	MANURE ONLY	MANURE AND BEDDING
1	100 lbs (45 kg)	200 lbs (90 kg)
2	200 lbs (90 kg)	400 lbs (180 kg)
5	500 lbs (225 kg)	1,000 lbs (450 kg)
10	1,000 lbs (450 kg)	2,000 lbs (900 kg)
20	2,000 lbs (900 kg)	4,000 lbs (1,800 kg)
30	3,000 lbs (1,350 kg)	6,000 lbs (2,700 kg)
40	4,000 lbs (1,800 kg)	8,000 lbs (3,600 kg)

Assumes one pound (0.45 kg) of worms eats ½ pound (0.23 kg) of material every 24 hours. Assumes three doublings per year (every four months is one doubling). Requires that normal weekly volumes are supplied and not excess from stockpiles.

Source: Colorado State University

U turn and advance back in the opposite direction, parallel to the first windrow.

OPTION 2. Create a base layer as you would in Option 1 that is 18 inches (46 cm) tall, 6 feet (1.8 m) wide, and as long as six weekly volumes of material will allow. Water the material and distribute *Eisenia* evenly across this 18-inch-tall windrow, as you would in Option 1.

Allow the worms to colonize this windrow and digest most of the material in the base. The time required for this will be dependent on the number of *Eisenia*. Monitor weekly for moisture and digestion.

Once the worms have digested most of the base layer, add a 3-inch (7.6 cm) layer of material on top of the base layer down the length of the windrow. Continue weekly monitoring for moisture and digestion. Create additional 3-inch layers at a frequency determined by how quickly *Eisenia* digest the material.

After this first windrow reaches 3 feet (0.9 m) in height, begin building the second base layer parallel to and touching the first windrow, as you did initially. As the worms begin colonizing this new base layer, add your first 3-inch layer to this second windrow. Add more layers as you notice *Eisenia* digesting the material.

Worms at Work

The *Eisenia* will travel from one windrow to another based on the moisture and temperature levels. Once one windrow begins to dry out and cool down, they will move to another one. Using a compost thermometer, make sure that the temperature stays below 90°F (32°C). If it rises above this temperature, you can make holes along the top to cool down the core and release heat. You will need to find a balance in the amount of water you give the windrow, as too much will cause air and worms to be forced out.

During colder months you may need to cover the windrow with a clear or black plastic sheet, though testing by CSU shows that these hardy little worms can withstand near freezing conditions. Leave the lower foot of the windrow exposed to allow for air circulation.

CHOOSE YOUR METHOD

To determine which method of making a vermicompost windrow, of the two described above, works best for your operation, consider the following:

OPTION 1

Advantages
- Involves less hand labor
- *Eisenia* population grows more rapidly
- Pathogens and weed seeds reduced

Disadvantages
- Digestion is not as thorough
- More likely that the windrow will become dangerously hot

OPTION 2

Advantages
- Less likely that the windrow will become too hot
- Digestion is more complete — more castings

Disadvantages
- More hand labor involved
- *Eisenia* population does not grow as rapidly

Before you harvest the vermicompost, allow several weeks for the worms to migrate from one windrow to another. Also, if you use cedar bedding, you will need to be sure that no more than one-quarter of the compost is composed of cedar bedding because it does not decompose very well due to its low appeal to microbial organisms.

MARKET THAT MANURE!

When properly composted, horse manure is a viable product that other industries and hobbyists love because of its fertile nature. There are two avenues you can pursue to market your manure: giving it away for free just so you don't have to move it yourself; and selling it to make a profit. For the following markets you can do either; which one you choose will depend on how much effort you want to put into the work and whether you want a part-time compost job. While you could have a side job just packaging the composted manure, you can get rid of it by offering a shovel-it-yourself option. Alternatively, you could find a young entrepreneur who wants to do the work and pay you a small fee for renting space on your property.

FARMERS AND RANCHERS. Those who aren't composting their own manure but are growing crops or other food are very interested in using well-composted manure.

GARDENERS. Rose gardeners in particular like the nutrient composition in horse manure. They prefer composted manure so that fewer weeds might grow in their gardens. Talk to local garden centers to do a cost comparison with current composts on the market.

MUSHROOM FARMERS. This is a great option for larger horse operations to market to, as these companies tend to value large amounts and a dependable supply. If you live in a county with a high horse population, you could consider co-oping the manure with your neighbors.

Consistent composting will be the key; mushroom farms will want a consistent nutrient level combined with a minimum amount of horsehair and bedding.

WORM FARMERS. Worms love compost. Fishermen love worms. Find a worm farmer in your area who will scoop up that compost for his worms.

Professional landscapers, organic farmers, and land-reclamation companies — businesses that deal in building and managing the land of others — are looking for sources of good compost. If you are not producing enough manure on your own property, you may be

It's Only Natural

Manure is a wonderful topic that always garners much attention from the non-horse public, specifically from spouses of horsepeople who just don't understand why so many horsepeople don't seem to mind the smell. EnviroHorse, a Web site based in California (see Resources), has quite a bit to say on the topic:

"Horse manure is recycled grass. Unlike a plastic bag, if you leave manure on the ground, it will quickly dry out and disappear with no intervention on the part of mankind. Society's debris and trash never go away and cannot recycle as can horse manure.

"Let's try to educate others to stop being so anal about horse manure. It is a clean, natural product used by landscapers, vineyards, farmers and big companies to provide topsoil, fertilizer and other gardening products. Poop is part of the natural cycle of life."

— California State Horsemen
EnviroHorse Web site

able to find another company already selling compost and combine yours. For example, feedlots often sell this by-product of their business. If the compost is not completely finished, it's often referred to as "mulch." Landscapers use mulch in their work.

Testing Your Manure

To market your manure effectively, you should have it tested for quality. Most university Extension offices offer manure-testing services for less than $50 per sample. They will test not only the carbon:nitrogen (C:N) ratio but also will analyze for nitrate, pH, electric conductivity, volatile solids, organic N, sulfur, zinc, sodium, chloride, calcium, magnesium, iron, manganese, copper, boron, and others, depending on the particular kit you purchase.

You will receive a report that lists the percentages for the minerals or nutrients that appear in your manure on a weight, volume, and dry-weight basis. (See the Resource section to find out how to download a sample report.)

Interpretation of the test is very important. Here are some tips:

- Compare the date you withdrew the sample, the date you sent it off, and the day it was tested. If there were any shipping delays, it may affect the outcome of the test.
- For comparison year to year, some labs will indicate what testing methodology was used. This will ensure that the same quality standards are used each year.
- Some values are actually measured, and others are calculated. For example: pH is measured; the C:N ratio is calculated; the total N is measured; organic N is calculated.

The manure and compost application rates are determined by the available nutrients. If you are selling your compost in an environmentally friendly manner, be sure that your client applies it properly to the destined crop

> ### When and Where Compost Should *Not* Go
>
> Do not apply compost to any land that is frozen or saturated. This will cause runoff of nutrients to water sources, such as streams, rivers, lakes, and gutters. According to Colorado State University, when excessive nutrients exist in surface waters, plant and algal growth becomes extreme, the oxygen supply is depleted, and fish can be killed.
>
> If sections of the land are highly erosion prone (such as riparian areas or hillsides), avoid placing any compost there, as it will be washed away.

or garden. If the nitrogen content is too high, it can cause nitrogen toxicity, which we'll discuss in chapter 11.

Testing is a key aspect of your compost-marketing strategy. Responsible landowners will be more willing to buy compost that has been tested and proven to be high quality. They are also responsible for their own recordkeeping, which will detail how much compost has been applied to their land and at what concentrations.

BEDDING CHOICES AND CONSIDERATIONS

The bedding you choose and the design of your stall floor are directly connected to the efficiency of your manure management system.

Starting with Straw

According to Alex Aballo, project director for Equestrian Services LLC, the original bedding of 50 or 60 years ago was mostly straw.

"Straw was used everywhere because it was inexpensive and widely available," she explains.

"There are several issues with straw, however. It's a very environmentally friendly bedding to use if you have mushroom farmers in your area that will come and haul it away and grow mushrooms on it. If you don't, there are the environmental disadvantages of straw. It is a lot more bulky than other bedding types, which means a lot more waste, and it does not compost easily because it has a waxy, water-repellent nature, so it doesn't break down as easily as other bedding."

Another reason not to use straw is the effect it has on your flooring. Its water-repellent coating means it doesn't absorb urine very well, so the urine travels right through your straw and into your flooring. If you have rubber mats (which are great, especially when recycled) but you are using straw, the urine will puddle under the straw and eventually degrade your mats.

Aballo suggests planning ahead and researching to see if you have enough mushroom farmers in your area who are willing to come and pick up straw. Find out how they will pick up, how often, how they want their waste, and the ratio of manure to straw that they would like in the waste they're going to use. If the demand is not there, you should seek other bedding options.

Shavings

Shavings are ideal, especially for composting, but the type of facility you possess will have an effect on the quality of your compost. If you are a private farm or even a lesson or boarding farm where, on a daily basis, you take out just the manure and urine-soaked bedding, then you have a much higher ratio of manure and waste to bedding, which is good. In this case shavings may be a good choice.

It's different, however, if you are a show venue or have many horses coming through your facility regularly. Every time a horse leaves your facility that stall needs to be stripped bare, because you don't want the next horse picking up diseases from the previous equine occupant. In this situation you will have a higher ratio of carbon to nitrogen because of the greater proportion of shavings in your compost pile.

Wood Pellets

Another option that is growing in popularity is wood pellets, a wood by-product (like shavings and sawdust), with the water and oils extracted. Because the product is compacted and compressed, the pellets occupy less space for storage. When you bed the stall, you moisten them, and they break down into a fluffy wood product. The pellets also are very effective at absorbing urine, so you remove more urine and manure but less bedding is going to waste.

The wood pellets come in hermetically sealed bags that are quite easy to manage. They can be stacked just like bales and should be kept under a tarp or shelter in case a tear in the bag allows water in (don't forget, they expand!).

☼ ❄ CLIMATE VARIATIONS

No matter where you go on this continent, it's good to know that some things will never change: manure will be the same manure wherever you go. And, like most constants in life, there will always be some commonalities in how we deal with it.

Subarctic

To ensure proper compost application in the subarctic climate zone, you will have to store a lot of manure over the winter season. If you are selling your compost, put extra effort into marketing for the short growing season so you can get rid of what you've composted during the previous winter and do not run out of storage room for the coming winter. Remember that you cannot apply manure to frozen pastures, because when the ground thaws in spring, your nutrients will quickly wash away into the nearest lake or stream. You can, however, spread any fully composted manure on your land (in an appropriate amount) prior to the first large snowfall of the year.

Humid Continental

Because of the weather patterns in the humid continental zone, you will need to keep a watchful eye on the forecast when you are considering applying compost to your fields. You will need several dry days in a row to ensure that the compost and its nutrients stay where you want them.

Humid Oceanic

The higher level of humidity in the humid oceanic zone means that you won't have to wet down your compost pile too much. In fact, you may have a harder time keeping it dry, so make sure it's tarped and does not have a chance to wash away in a rainstorm.

Highlands

It can be a challenge to spread manure effectively in higher elevations. If you live in this zone, choose level spots for manure bins or windrows to ensure nutrients are not washed away. Consider selling as much of the compost as you can. For example, tree farms in your area may require more compost because of the rockier nature of the ground.

Semiarid

Although your risk for manure fire in the semiarid zone is not as great as some other zones, you should still take precautions during periods of drought. You are in a region that has a high marketability for compost, as the soil in your region can be depleted quite easily from overgrazing.

Arid

The risk of a manure-pile fire is greatest in the arid climate region. Make sure that a reliable water source is located near your manure pile, and keep a watchful eye on the moisture content.

WATER, WATER EVERYWHERE

How to evaluate your needs and maintain

and conserve clean water

WATER IS the single most important resource that any and every living creature needs — and fortunately Earth, "the water planet," has plenty of it. The most pristine, beautiful land in the country, however, can be useless and even dangerous if the water becomes contaminated. In this chapter we'll discuss how we can make the most of what we have, in both quality and quantity.

FLOWING WATER

How can you find out how safe your flowing water is? Here are three places to start.

First, investigate what is located upstream from you, because if you have water running through your property, you'll want to know who is putting what in it before it reaches you. For example, there may be a larger city upstream, and you should know what its water-treatment policy entails. What do they allow to be dumped into the waterway, and how do they notify residents of contamination?

Second, look to your neighbors' property. Do your neighbors use chemical fertilizers or store their manure improperly? As you consider your and your neighbors' environmental impact, it is important that you understand the layout and grading of your own land. If your land is just a few feet lower than your neighbors', any runoff will come straight down to your property and possibly into your water system. Even flat land can share the same groundwater supply.

Third, check your own property. You should learn not only where the water on your land originates but also where it goes when it leaves. Do you know where the water flows on your land, both above and below the ground? If you have a manure pile that you are using for compost, is it impeding or polluting water flow? If your pile is not protected from rain by a roof or tarp, not only could you be contaminating water around your property, but the nutrients that make that manure great compost material are being washed away. There are many different possible contamination sources on your own property.

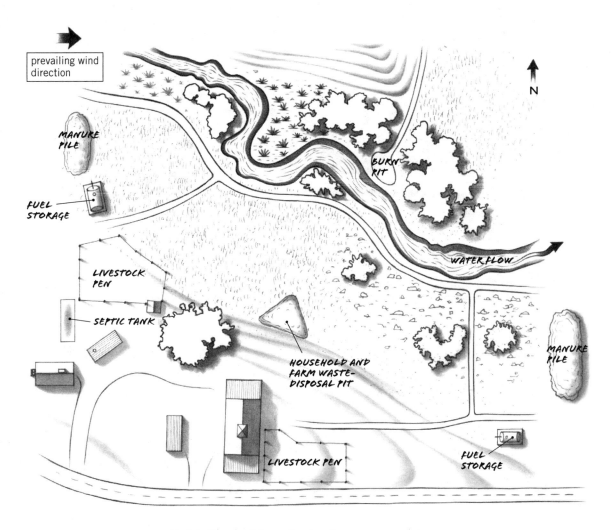

Sketch the possible contaminants on your property.

Common property contaminants include:
- Household and farm waste-disposal pit
- Sewage system
- Fuel and chemical (pesticide or fertilizer) storage location
- Livestock pens
- Manure pile
- Burn pits (largest source of dioxins)

These contaminants can be enough to deal with when contained in one location, but water, the universal solvent, distributes and disperses any such elements it encounters. The most common problem occurs when runoff from buildings is unchecked during heavy downpours and spring snowmelts. Minimize the amount of unchecked water flow on top of your land by using a roof-water catchment system (see chapter 5).

WELL WATER

Many landowners (especially those with land located outside city limits) use well water for their homes, barns, and automatic waterers. This means that the landowner has sole responsibility for maintaining the water system and keeping the water free from chemicals and microbiological hazards. This is a great responsibility and not a small undertaking.

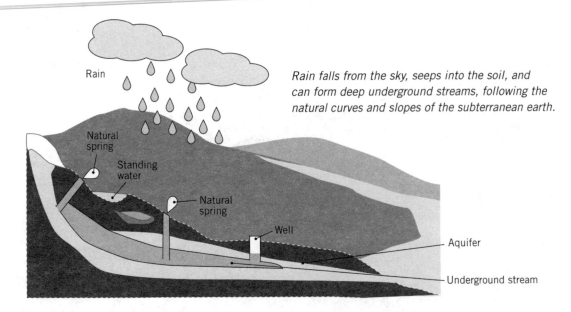

Rain falls from the sky, seeps into the soil, and can form deep underground streams, following the natural curves and slopes of the subterranean earth.

How Does Good Well Water Turn Bad?

According to the Canadian Water and Wastewater Association, wells become contaminated in many ways:

- Ground spills (example: fuel and pesticide spills)
- Injection into the ground (examples: septic leaching beds, disposal of waste in wells, contaminated surface water running into poorly constructed wells, poorly maintained wells, improperly plugged wells, and back siphoning from spray tanks into wells)
- Improper handling of industrial solvents and chemicals (examples: Varsol and wood preservatives)
- Waste leakages (examples: manure storages, wastewater, septic tanks, and landfills)
- Underground and aboveground fuel-storage tanks that leak
- Movement of groundwater between contaminated and clean aquifers
- Overapplication of manure, commercial fertilizers, or pesticides

Any change in the taste, appearance, or smell of well water should be investigated immediately. Take note if horses start to turn their noses up at water that appears perfectly fine. Look also for scaling in sinks and faucets, discoloration of laundry or plumbing fixtures, and any unusual way dish soap reacts with your water. These could all be signs of contamination.

Testing the Waters

Regular testing of your well water is crucial to ensure that your water is clean and safe. There are two types of testing:

BACTERIAL. A bacterial analysis should be done annually to test for the two organisms that indicate the presence of disease-causing bacteria: coliform and fecal coliform bacteria. This analysis does not test for iron bacteria or sulfate-reducing bacteria, both commonly found in well water and safe for consumption.

CHEMICAL. Every three to five years you should have a chemical analysis done on your water, which measures the levels of iron, sodium, sulfates, nitrates, and nitrites. If a health concern has been raised in your region because of an increased potential for contamination (such as a chemical spill), you will need to request testing for the specific substances.

An aquifer is permeable geologic material through which large amounts of water can move underground. It includes unconsolidated material, such as sand and gravel; permeable sedimentary bedrock, such as sandstone, limestone, and dolomite; and fractured crystalline bedrock.

Water does not flow through an aquifer the way it does through an aboveground stream. In fact, it might only move a few inches a month.

Finding an aquifer on your property is a job for a surveyor or your local Cooperative Extension Service. If a well is already located on your property, you may also be able to get this information from the company that drilled it.

COLLECTING A WATER SAMPLE

When you collect a sample of well water, take it as close to the wellhead as possible. If you collect the sample from inside your house or barn, your water treatment process will skew the results.

The pump should run at least 10 minutes before the sample is taken to ensure anything inside the pipes is flushed out. Make note of the length of time the well is pumped prior to the sample retrieval, how the sample is stored, and the time it takes to deliver the sample to a laboratory.

STANDING WATER: PONDS AND LAKES

Even barring environmental or chemical contamination, standing water in lakes, ponds, or dugouts on your property still can become polluted from misuse by careless property owners. The types of pollution that affect standing water are known as primary and secondary pollution.

Primary pollution occurs when organic material decomposes in your water, consuming the dissolved oxygen that is necessary for fish and aquatic animals. This organic material most often comes from paddock or manure-pile runoff.

Secondary pollution occurs when the water system becomes a breeding ground for plant species such as algae, which are fertilized by the nutrient-rich runoff. These plants begin a growth-death-decomposition cycle that claims even more oxygen from the water.

Keeping standing water clean and free of organic waste ensures that the water your horses drink is clean and healthy. You are also supporting a necessary ecosystem on your property — healthy water means healthy animal and plant life.

PROTECTING YOUR WATER

The three types of water on your property — well, standing, and ground — can be protected and maintained through good land stewardship.

Unless you are building from scratch, you may not have any choice about the land where your barn rests. Be it hilly, flat, or a combination, you will need to understand the lay of your land so you can determine the directional flow of water.

Diversionary Tactics

There is not much you can easily do to change the flow of groundwater, but standing water or water that flows through your property can be protected from runoff by encouraging natural barriers and constructing man-made berms or berths. Trees and shrubs planted along waterways are examples of natural barriers; man-made structures might include ditches or long troughs dug to direct runoff elsewhere.

During rainstorms or spring snowmelts, watch where your water flows. If there are low spots where the water gathers, you can try to divert it either before it reaches that spot or once it has reached the low spot. The latter can be more difficult because you may have to dig a culvert or create underground drainage to move the water away or promote infiltration. A safe direction may be toward pastureland located away from your water supply or into ditches designed specifically to catch water. (There is more to say about drainage later in the chapter.)

Smaller properties may not have the luxury of pastureland to which they can divert the water. In these cases, digging a series of smaller ditches to divert the water and spread it out over a greater area might be your best bet. Tips for doing this will be discussed later in this chapter in the section Improving Drainage in High-Traffic Areas.

Planning to Avoid Pollution

The greater goal is to keep any possible contaminants away from the water in the first place. Locate your manure pile, fuel tanks, storage facilities, and other possible contaminant sources away from any source of water. Check with your local zoning board to determine if there are specific by-laws about the setbacks required between possible contaminants and water supplies.

When building manure piles, fuel tanks, and storage facilities, place impermeable barriers between them and the ground. These can be as simple as a cement pad that catches any spillage. Identify the direction of the grade on all

---

One in a Million

According to the U.S. Environmental Protection Agency, a 1-gallon (4 L) fuel spill can potentially contaminate millions of gallons of water.

---

four sides of these areas, and monitor for any signs of contamination down the grade, such as dying grass, discoloration of ground or plants, increased growth in a distinct pattern, or saturated ground. Also, covering these sources can prevent rain from washing contaminants away.

According to the 2000 National Water Quality Inventory, agricultural pollution of waterways resulted from "poorly located or managed animal feeding operations; overgrazing; plowing too often or at the wrong time; and improper, excessive or poorly timed application of pesticides, irrigation water, and fertilizer" (EPA 841-F-05-001, *Protecting Water Quality from Agricultural Runoff*). Pollutants include sediment, nutrients, pathogens, pesticides, metals, and salts.

Sediment refers to soil that is washed away from agricultural land. These soil particles can cloud water (reducing the amount of sunlight that reaches aquatic plants), clog the gills of fish, and even harm fish larvae. In addition, sediment often brings fertilizers, pesticides, and other pollutants along for the ride. Combating sedimentation is a matter of minimizing runoff while using land-management practices to keep the soil in place.

Nutrients are often applied to land in the form of chemical fertilizers or manure. Since we are choosing nonchemical horsekeeping practices, we will focus on nutrients in the form of manure. As mentioned above, nutrients washed into waterways can cause many problems. This is why you need to be careful when applying compost to your fields (as outlined in chapter 7) and keeping manure and compost piles safe from runoff and protected from rain saturation.

Overgrazing not only damages the land, but it also contributes to water contamination, especially sedimentation, by destroying the plants and root systems that keep the soil where it is supposed to be. By implementing pasture- and grazing-rotation programs, you can minimize overgrazing.

Dirt Glue

Dirt Glue Enterprises has created an environmentally friendly, nontoxic, water-based acrylic polymer emulsion that can be placed on any agricultural surface to stabilize soil in areas prone to erosion. It also has applications for dust control on roadways and inside arenas. Dirt Glue has been used by the U.S. military to control dust in critical sight areas and has been used in hydroseeding operations to speed the process of germination by holding seeds in place.

(See Resources for contact information.)

Some smaller farms may not be able to rotate pasture space that is already at a premium. In this case you have several alternatives:

- Place several water and shade sources around the pasture to encourage animals to move to different areas
- Decrease intensity of use by reducing the amount of time horses graze the pasture each day
- Fence off sensitive areas or portions of the field
- Decrease the number of horses

This topic will be discussed at greater length in chapter 9.

THE WATERSHED

When it comes to water, everything is connected. Snowfall, spring thaw, rainfall — it all comes from the sky onto the ground and into the rivers, runs to the sea, then evaporates into the clouds and falls again on the ground and the rivers. Those areas of land where the melting snow and rainwater gather are called watersheds. From the watershed, runoff flows or seeps into surface waters such as lakes, rivers, and streams, as well as into groundwater.

Each watershed area is unique and has its own geography — some of it natural and some man-made. Because you spent time making your land maps in chapter 3, you should have a good idea of where any watersheds are located on your property. Keep a close watch on the health of the grass, and ensure that your yearly soil testing covers this area. The health of the watershed area is an indication of the overall health of your property.

Riparian Areas

A riparian area is biodiverse because it features the perfect trifecta of water, vegetation, and rich soil that drives the cycle of life. Healthy riparian zones reduce soil erosion, act as a filter for water entering the system, and reduce the effects of drought by retaining water and releasing it slowly.

If you have standing, flowing, or surface water on your property, the riparian areas that surround them are your responsibility. These areas are particularly vulnerable to grazing animals, and you must take care to monitor these areas for signs of damage (see box on page 125).

- Damage to trees, young or old, by breakage or damage to bark
- Overgrazing or shortened grass
- Animal tracks left in soft ground (evidence that roots are being damaged)

GRAZING GUIDELINES

Lush riparian areas should be used judiciously for grazing so you strike a balance between conservation and practical function. Following are some suggestions.

DEFINITION: Riparian Area

The area directly around a watercourse.

- If you are going to put animals to pasture in this area, avoid the springtime, when many plants and animals are at the start of their life cycles and are most vulnerable.
- Schedule shorter grazing periods to prevent damage.
- Encourage horses to graze away from riparian areas by placing water, salt, and feed elsewhere to reduce the risk of water contamination.
- Don't use the stream as your only water source.

Many counties and municipal districts in North America require that you fence off riparian areas from livestock — important even without regulation. Depending on your region, government programs may help pay for this.

Check with your local office to make sure you are following the correct guidelines and regulations regarding fencing and buffer zones. Many horse owners mistakenly believe that creeks on their property are there to water their horses so that they don't have to.

Alex Aballo, project director for Equestrian Services LLC, has had direct experience with protecting natural streams on her property. Two creeks ran through her farm and ended up in a large reservoir. When it came time to fence her pastures, she had county soil conservation staff come out and tell her how much space she needed to reserve as a buffer for the creek.

"I had to fence the horses away from the water source," she said, "so that manure would not leach into the water and the horses wouldn't traipse through the water and contaminate it."

THE WATER CYCLE

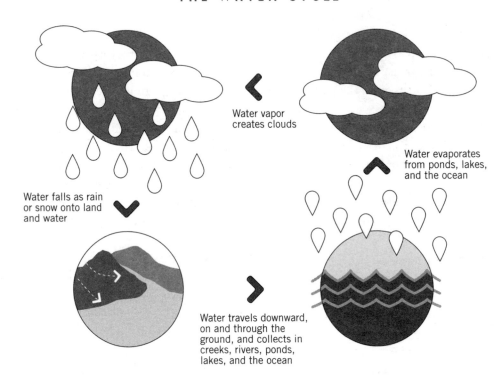

Water vapor creates clouds

Water evaporates from ponds, lakes, and the ocean

Water falls as rain or snow onto land and water

Water travels downward, on and through the ground, and collects in creeks, rivers, ponds, lakes, and the ocean

An Expert's View: Preventing Erosion on Stream Banks

Q: *Are there plants that will grow in my pasture to help prevent soil erosion? I have several creeks running through my property, and the banks seem to crumble more each spring.*

A: All grasses do a good job of reducing soil erosion, but the creeping rooted or rhizomatous grasses may be the best choices. Grasses such as creeping red fescue, smooth bromegrass, or Kentucky bluegrass would be very effective at preventing soil erosion alongside creeks. The grasses growing in these areas generally grow more forage and remain green longer because they have access to more moisture than the upland grasses. Because of this, the livestock prefer to graze and loiter in these areas, which creates a problem of overgrazing.

When grasses are always overgrazed, they become less productive, as the root system is providing energy to produce the top growth. The lack of top growth does not allow the plant to use the sun's energy through photosynthesis to provide and return energy to the roots. The roots then become shorter and weaker, which affects the soil-holding capacity of the roots, and erosion occurs.

If possible, these riparian areas should be fenced to control livestock access to these grasses and to the creek itself. The livestock should only be given access to the creek at graveled access sites, where there is less chance of erosion. In many cases this is not possible, so you should look for access sites where there is less chance of erosion. Ideally, some type of stock-watering system should be developed to keep cattle and horses out of the creek and away from the creek, but this can be expensive and may not be practical.

It may take more than one grazing season to rejuvenate these riparian areas, but once they become productive again, some type of controlled grazing is possible as long as the grass is allowed sufficient rest periods to maintain a strong, deep, fibrous root system. Planting willows along the creek will also aid in reducing erosion.

It is not the type of grass as much as its management that is causing the erosion. It is the "too soon, too long, too much, and too often" grazing management of riparian areas that leads to erosion.

You do have to be careful not to plant grasses that are foreign to the area, and keep in mind that introducing any plant near a stream means you are taking a risk that the seeds could possibly get into the waterway and be introduced to your neighbor's land as well.

— Arvid Aasen, Airdrie, Alberta, Canada, is a forage and pasture agronomist with the Western Forage/Beef Group at Alberta Agriculture, Food and Rural Development

Native grasses are one of the best methods to prevent streamside erosion.

GETTING WATER TO YOUR BARN

Many older facilities do not have adequate water access to their barns. Often this is because digging trenches, putting in pipe, and relandscaping can cost a lot of money. But over the long term, piping water directly to your barn saves the environment because you're not wasting the water that drips from hoses as you drag them to reach your barn. If you have more than a few horses, you're not wasting water by hauling it in buckets, and importantly, you're saving labor — and your back.

Of course, many farms use older trucks that are dedicated to being "the water truck" and have tanks on the back specifically for hauling water. Great idea — except that using an old (and probably gas-guzzling) truck to move water around your property may very well cost you in the long run. Think about the amount of fuel you use each month, factor in the environmental cost, and ask yourself if you can afford to pipe water around your property instead. It's likely that you can.

An easy source of available water right near your barn is the water runoff collected from your roof. As we discussed in chapter 5, this is the best way to get water to your barn without expensive piping. The water from your catchment can be fed to your horses as long as it has been treated; untreated, it can be used for irrigation.

CONSERVATIVE WATER USE

Horses are natural water conservationists. Not many horses will use more water than they need to survive. They don't choose to take longer showers or request that their water have ice in it on a hot summer day. As their caretakers it's up to us to provide an adequate amount of clean, safe water for them. Being conservative with your water use does not mean limiting water for your horses.

Need Help?

The best resources in water and watershed management are available at your local or regional level. Local agencies will have specific information that can help you manage your water relative to your region's challenges and opportunities. A variety of federal agencies can help as well. (See Resources for Web addresses.)

A horse who is not doing much of anything still requires 8 to 10 gallons (30 to 38 L) of water a day, while horses in training or who are pregnant or lactating may need closer to 15 gallons (57 L). Water is crucial to a horse's health, just as it is to ours — and perhaps even more so because it provides lubrication for his lengthy digestive system. (Ours has the benefit of gravity to keep things moving — his needs to push digested food sideways!)

There are many ways water can be lost on the farm. The two main causes of water loss are accidental spillage (by horse or human) and evaporation.

Six Ways to Water Your Horse

Here are some guidelines for watering horses efficiently.

1. **Use rainwater safely gathered** and kept free from debris. You can use the runoff from your barn to water the turnout paddocks closest to the barn, making sure that the water troughs and buckets are clean and filled regularly. A simple filtration and UV-treatment system can be installed right on the side of your barn, and if you install the cistern or water barrel higher than the paddocks, you can use gravity to fill the troughs.
2. **Use automatic waterers** in your paddocks and pastures that are located on higher ground.

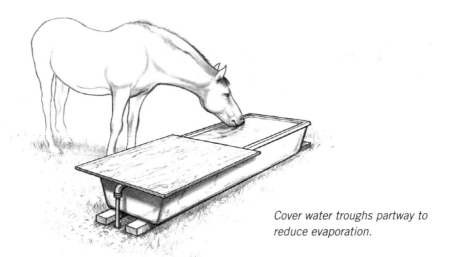

Cover water troughs partway to reduce evaporation.

This ensures that any runoff will water the pasture and not gather in a low spot to create a mudhole.

3. **Keep water troughs mostly covered** and in cool, shaded locations to cut down on evaporation.

4. **Fill water troughs daily,** and keep them clean so the horses drink the largest amount possible and never encounter dirty or scummy water, which will stagnate and be wasted.

5. **Insulate your troughs** to keep the water warm during wintertime. Horses do not always like cold water and will drink less, which leads to stagnation or evaporation rather than good use inside your horse's belly.

6. **Install a small solar-powered water heater** or circulator to keep water moving (and therefore not freezing) so your horses will drink more.

Put Gray Water to Work

A lot of water waste can occur inside the house and barn on your property because of improperly maintained or monitored facilities. The toilet that runs and isn't noticed right away; leaving the hose running while bathing a horse; or an indoor waterer with a stuck float: these are common water-waste culprits. Dripping faucets, waterers, toilets, and hose connections

---

Water Concerns for the Boarding Stable

If you have boarders, it's very important to outline water use and conservation strategies to ensure that they are complying with your rules about how and when to use water. Boarders who use water frivolously can end up costing you more money than they are worth. Consider posting rules on a common bulletin board. Here are some ideas:

- If you turn it on, turn it off
- Bathe horses with buckets and sponges only and only in approved locations
- Report any drips or leaks
- Do not fill water buckets yourself but report empty buckets to management

---

should be fixed immediately. Repairing leaks saves water and money in the long run.

Here is an excellent water-saving choice: convert your barn and household bathroom facilities to use gray water. Gray water is shower, sink, and laundry water that has been filtered and disinfected (with chlorine or other purifier) before going to your toilet bowl. In busy boarding facilities ten or more people may use the bathroom in a day, and not many nonfamily

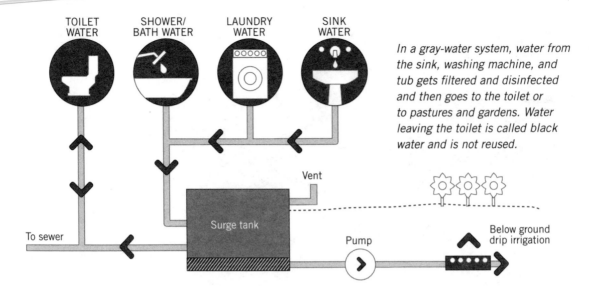

TOILET WATER   SHOWER/BATH WATER   LAUNDRY WATER   SINK WATER

Vent

To sewer

Surge tank

Pump

Below ground drip irrigation

In a gray-water system, water from the sink, washing machine, and tub gets filtered and disinfected and then goes to the toilet or to pastures and gardens. Water leaving the toilet is called black water and is not reused.

members are comfortable obeying the "if it's yellow, let it mellow — if it's brown, flush it down" motto. Smaller farms, though, can often do without a bathroom at the barn because the house is a stone's throw away.

Reuse is not a viable option, however, for the water that comes out of your barn washracks because it is too contaminated to be recycled into gray water. It is considered black water because of the high manure and bacteria content.

TOILET TALK

Choosing a low-flow or dual toilet at home or barn is a great way to conserve water. Low-flow toilets can use as little as 1.6 gallons (6 L) per flush (GPF), quite a difference from the older versions that use up to 7 gallons (26.5 L) in a single flush. Dual-flush toilets offer two settings to choose from, based on the needs per flush: 1.6 GPF or 0.8 GPF (3 L).

Another option is using an incinerator toilet. These often run on electricity, however, costing you more money if you are not using alternative-energy sources. Incinolet is a popular brand that has been in use for more than 40 years. The Incinolet uses 120 volts (there are some 240-volt models) at 15 amps, which translates to about 1½ kilowatt hours of electricity per cycle. Electricity is used only when the toilet is in use.

Waste is incinerated after each use, and no additives are required. The ash produced must be removed each week and disposed of along with the regular garbage. It is completely clean with no surviving germs or microbes. (See Resources for contact information.)

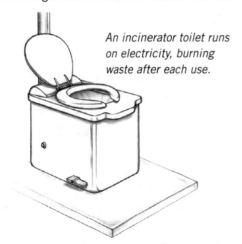

An incinerator toilet runs on electricity, burning waste after each use.

Monitor Automatic Waterers

It's important to ensure that all pipes and connections to your automatic waterers are sealed and free from leaks. A constantly dripping waterer will waste hundreds of gallons of water each year. Outdoor waterers must also be checked frequently for stuck floats and leaking pipes. Know exactly where the water pipes run underground and look for wet, soggy areas that might indicate an underground leak. You can also purchase livestock waterers that offer constant circulation (powered by solar or wind) to discourage algae growth in the summer and freezing in the winter.

Some horses may need to learn how to drink from waterers that have a large ball or float that seals against the tank and must be pushed down to produce a drink. These waterers are generally insulated against freezing during the winter and protected against evaporation in the summer.

Cover and Maintain Large Water Tubs

If you water your horses using large troughs or tubs, cover three-quarters of the tub with boards or plywood to minimize evaporation and discourage algae growth. Check for algae growth every two or three days. You can add an ounce or two (30 or 60 mL) of hydrogen peroxide to 100 gallons (380 L) of water to keep algae growth at bay. Horses may turn away from water that has an abundance of algae. If you find yourself battling algae again and again, you may need to scrub the tub clean. Using a small capful of chlorine to 100 gallons of water can also help inhibit algae growth.

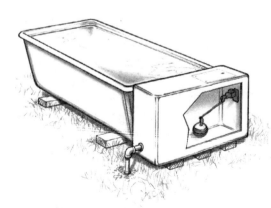

MUDDY PADDOCKS

Boarding stables and barns with heavy traffic leave a large environmental footprint on the land. Turnout paddocks are often small and close to the barn, meaning the paddock and barn areas bear heavier human and horse foot

If you have horses who love to play in the water, they can turn the area surrounding their waterer into a shoe-sucking muddy mire. Set the tub or waterer on boards or a concrete pad for a safer, tidier setup and ensure everything is located in a low-traffic area with good drainage.

traffic, which does not allow for much grass growth.

Outdoor paddocks where horses are kept year-round will also degenerate into dirt if they are not large enough for the horses they house, or if they are not rotated consistently.

Your Anti-Mud Program

Encouraging more grass in your pastures and paddocks will not only cut down on feed usage but also will keep the enclosures from becoming mud holes during wet seasons. Alayne Renée Blickle of Maple Valley, Washington, a lifelong equestrian and reining competitor, is the creator and director of Horses for Clean Water. She provides eight key points to reduce mud:

1. **Create a sacrifice area** or paddock to use during the winter. Locate it on higher ground and away from creeks, wetlands, ponds, or other water bodies. Surround this paddock with a grassy area; the vegetation will provide a filter for contaminated runoff. Keeping horses in a sacrifice area prevents them from destroying pastures and turning them into a mud mess. It also confines manure to an area where you can pick it up easily and compost it. That messy, muddy paddock horses create by mixing melting snow into the dirt translates to a very dusty paddock in the spring and summer. Remember, winter mud = summer dust.

2. **Add some type of footing** in sacrifice and high-traffic areas. A footing material will cut down on mud problems by keeping horses off the dirt and allowing rainwater to percolate through. Popular choices include hog fuel (chipped wood), gravel (crushed rock), or sand (see box on page 134). Hog fuel has the added benefit of helping break down the nitrogen in the horse's urine and manure. Use 3 to 6 inches (8–15 cm) of footing throughout your paddock, depending on how deep your mud is.

3. **Pick up manure** in your paddocks and sacrifice area every one to three days. This is important for your horse's health because it reduces parasite reinfestation, and it will also greatly decrease the buildup of mud. While you're at it, pick up stray clumps of bedding or leftover hay. All organic material eventually decomposes and fuels mud.

4. **Tarp the manure pile.** This will keep it looking like a pile of compost and not a pile of mush. The nutrients you are trying to save will remain in the compost and will not wash out into surface waters where they can cause problems. Store manure as far away as possible from streams, ditches, and wetlands to avoid mud problems and potential environmental impacts.

5. **Manage surface flows.** If surface flows run into your barn or paddocks, look for ways to divert this water. Possibilities for dealing with surface-water drainage include French drain lines, water bars (like a speed bump for water runoff), rain gardens, infiltration wetlands and ponds, grassy swales, and dry wells.

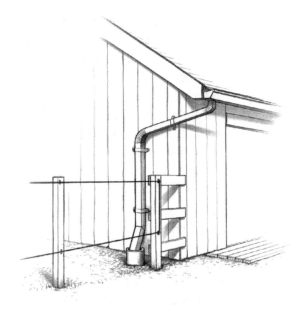

Divert rainwater away from high-traffic areas with gutters and downspouts.

6. **Install rain gutters** and downspouts on all barns, sheds, and outbuildings, and divert clean rainwater away from high-traffic areas. A 12-foot by 14-foot (3.7 × 4.3 m) shed can produce as much as 6,000 to 12,000 gallons (23,000–45,000 L) of rainwater in one year in the Northwest. Diverting this clean rainwater away from your paddocks and high-traffic areas will substantially reduce the amount of mud created. This has the added environmental benefit of reducing the amount of nutrients (from manure) and sediments (from mud) that get washed into surface or groundwater. (See chapter 5 for more on roof runoff catchment systems.)

7. **Keep horses off rain-soaked soils** and dormant pasture in the winter and early spring. This is critical if you want to avoid mud this winter (and dust next summer) and if you want a healthy pasture next summer. Soggy soils and dormant plants simply cannot survive continuous grazing and trampling in winter months. Horses are particularly hard on pastures — the pounding of their hooves compacts the soil and suffocates plant roots. In addition, when soils are wet, horse hooves act like plungers by loosening fine particles of topsoil that are then washed away by the rain, creating mud and other environmental concerns.

8. **Plant and maintain native trees and shrubs.** Plants use a lot of water and can potentially reduce the amount of water around your horse place. A mature Douglas fir can drink 100 to 250 gallons (380–950 L) of water per day. Evergreens have an added advantage in that they continue to use water in the winter when deciduous trees are dormant. Planting water-loving native shrubs along the outside of paddocks may help keep the area dryer, and it will reduce runoff. Examples of plants that might work include willow, cottonwood, red osier dogwood, and hybrid cottonwood.

Native shrubs and trees (such as these cottonwoods) will drink up a large quantity of your excess water and help reduce mud.

Pristine Paddocks

If your outdoor turnout pens tend to get muddy, to the extent that your horses need to stay inside rather than be turned out, add cedar shavings or chips as paddock footing to reduce mud buildup. Choose midwestern aromatic cedar rather than northwestern red cedar, as the latter can be toxic. The rice-like textured cedar chips will not blow away or decompose. When wet, they can simply be turned over to dry in the sun and wind. Cedar was designed by nature with high levels of cellulose and lignin, which are not readily available as food to most microorganisms, so it will not decompose for years to come.

The aroma, while pleasing to humans, is a fly deterrent. And all you have to do is pick up the manure each day to maintain the quality of the product. If you don't pick up manure regularly, it will gradually mix into the bedding.

To install properly, start with a dry, firm, and level paddock with proper drainage, and simply spread the product evenly. You'll get a healthy alternative to fly sprays, minimal dust (you can wet it down, and it will dry on its own without blowing away), and comfortable and safe footing for your horse.

Improving Drainage in High-Traffic Areas

In every paddock a little mud will gather . . . or at least in improperly managed paddocks it will. Whether in daily-turnout pens, permanent paddocks, or pastures, there are always high-traffic areas where horses eat, drink water, or gather. Often these spots are found — logically — near automatic waterers or tubs, hay mangers or feeding areas, gates, and shelters.

Very often less planning goes into the selection of turnout paddocks because the only criterion for choosing their location is closeness to the barn for convenience. This is not a wrong decision, but you need to consider the extra maintenance it may involve. Alternatively, turnout areas may connect directly to each horse's stall. In this case it's vital that water from the barn roof be diverted correctly. Even when precautions are taken, paddocks still may be affected by runoff and standing water.

Factors that might exacerbate mud are sheds or shelters inside the paddock, downward-sloping or low-lying areas, and poor manure management in the paddock. Many horse owners believe that the reason to have large fields and no indoor stalls is simply so they don't have to pick up manure. Not so. As Alayne Blickle mentioned earlier, manure needs to be collected to avoid both flies and mud.

You may need to consider some drastic measures if there is an area that's particularly difficult to keep mud-free or a location that has historically been a mud pit. Managing water and mud go hand in hand. The basic premise with installing drainage improvements in areas like this is that you are putting the clean water to good use while allowing dirty water to be filtered through grassy areas or special drains.

DEFINITION: French Drain

Also called a drain tile or land drain, a French drain is a ditch filled with gravel and rock that redirects surface and groundwater away from a specific area. Very often it is used around the foundation of a house or barn. It is equally effective, however, in diverting water away from paddocks and pastures.

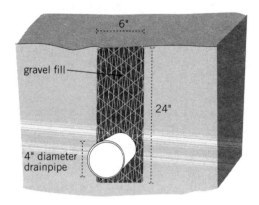

A French drain is a drainpipe embedded in a gravel-filled ditch to direct water away from a specific area.

CASE STUDY: RECLAIMING A MUDDY GATE AREA

The University of Vermont's Ellen A. Hardacre Equine Facility undertook a paddock renovation project called "Greener Pastures" in 2004, when sections of their paddocks became so muddy that it was difficult to bring horses in and out of gates. The university renovated the gate areas of the worst paddocks, and the change by the following year was so dramatic that the remaining paddocks were also renovated. In an article in the *Journal of Extension,* Elizabeth Greene, associate professor and Extension equine specialist at the University of Vermont, explains the process:

> We renovated a 15 × 76-foot (4.6 × 23 m) strip of land along the front of each paddock. Renovation involved replacing 8 inches (20.3 cm) of compacted topsoil with a layer of geotextile filter fabric, 4 inches (10.2 cm) of large stone (1½ to 1¾ inch [3.8 to 4.5 cm]), covered by another layer of fabric, then 4 inches of dirty pea stone on the topmost layer. The "large stone sandwich" allowed water to flow underneath the compacted top surface to a slightly angled PVC pipe buried under the travel lane to the grass buffer and into the existing French drain.

The width (into the paddocks) of the renovated area was determined by both the amount of area needed to lead a horse in and turn it around prior to releasing it, and the available width of the filter fabric roll. The entire length of the front of each paddock was renovated. The depth of each type of stone was approximately 4 inches, and, although the upper layer was hard pack, it was no harder than the previously compacted ground, and the mud and ice issues were significantly reduced.

The two layers of geotextile filter fabric were necessary to prevent both the soil (bottom) and the pea stone (top) from filling in the air pockets that allowed water to pass through to the drainage pipe. The university hired a contractor to do most of the work, because the farm labor and equipment was not available due to other farm commitments. As a result, each paddock (labor and product) cost approximately $1,400 dollars to renovate.

THREE OPTIONS FOR SACRIFICE-AREA FOOTING

$ — HOG FUEL

Hog fuel consists of large wood chips that need to be applied twice as deep as the depth of the mud, up to 12 inches (30.5 cm). It will pack down and degrade over time, and new material will need to be added each year. It also reduces the smell of urine.

$$ — GRAVEL OR SAND

Gravel and sand are about twice as expensive as hog fuel, but they last longer. The best size is ⅜ inch (0.95 cm) minus to ⅝ inch (1.6 cm) minus crushed gravel in a 2:1 ratio of gravel to mud. You can apply up to 8 inches (20 cm) of gravel. It eventually migrates into the soil and will need to be replaced. Coarse, washed sand is also an option but should not be used in areas where horses may be fed due to the potential for sand colic.

$$$ — GEOTEXTILE CLOTH WITH GRAVEL, SAND, OR HOG-FUEL FOOTING

You can put a water-permeable cloth beneath any of the footings mentioned above to prevent them from migrating into the soil while still providing drainage. You need to put the geotextile cloth down on a level, graded surface and cover it with a minimum of 6 inches (15 cm) of footing. If the area needs extensive drainage (for example, it's in a low-lying area), add at least a 3-inch (7.6 cm) layer of crushed rock, followed by a second layer of fabric, followed by the footing of your choice.

Source: *Managing Small-Acreage Horse Farms*, Oregon State University Extension Service

CREATING BUFFER ZONES

Buffer zones can protect an area from runoff from another area and should be placed between any water source and a possible contaminant area such as a manure pile or paddock. A buffer zone filters mud, manure, pesticides, and bacteria out of the water that flows through it and protects the land around it from erosion during wet seasons.

These areas can be used in your grazing rotation (for short periods of time) or can be used to grow a profitable crop. A common choice is to plant something that will not be consumed, such as shrubbery that can be used by landscapers and garden shops. Keep in mind that this area should not be fertilized, to avoid transferring nutrients to any other areas via the filtering water.

While placing smaller buffer strips between paddocks may prevent cross-contamination, a good-size buffer will be about 50 feet (15.2 m) wide to allow for adequate filtration of the water. The steeper the grade and the more fine grained or dense the soil, then the wider it should be.

If space is at a premium, you may consider creating a raised section of earth called a berm to protect sensitive waterways. This works best if the berm directs the water to a filtration system, such as a French drain. A French drain is most effective as a conveyance system.

Subarctic

Water, water everywhere . . . it's just frozen most of the time in the subarctic region. Your biggest challenges will be managing spring runoff and keeping your horses hydrated through the winter months. Water collection from rooftops will assist in this, as the snow in this region will begin to melt and then go through various stages of melting, freezing, being covered with more snow, and melting again. Make sure you keep an eye on ice buildup in gutters and spouts. Keep riparian areas fenced off, even in the winter when most areas are frozen, because as the thaw begins, horses can still damage these areas by stepping through ice along the sides of streams, damaging dormant root systems.

Humid Continental

You will experience a lot of snowfall in the humid continental region, especially if you are near the Great Lakes and receive "lake-effect snow" each year. Managing aboveground water flow during spring thaws will be very important, especially in paddocks and pastures that are located close to buildings.

Humid Oceanic

Because you have such a wide range of plant life in the humid oceanic climate zone, you can use many different grasses, legumes, and shrubs for erosion control on hillsides and in sensitive areas. Shrubs are often your best choice, since tasty legumes may encourage horses to venture into areas that can erode easily. Consider establishing temporary fencing that you can move weekly to give paddocks time to recover from rain; this is in conjunction with proper sacrifice-area management, which will be covered in chapter 9.

Highlands

In the highlands climate zone, you need to make sure that you know where the above- and belowground flow of water is occurring before the spring thaw. Pay close attention to the entire grade of your land when establishing water-management practices. It will always be easier to transport water from high to low grades, so make sure your system takes advantage of the land slope and gravity.

Semiarid

Water management in the semiarid region is very important. Riparian areas will need extra protection, since they have a challenging existence as it is. Fencing off water is very important, but make sure that you have enough water for your horses, and do all you can to minimize evaporation, including locating water sources (buckets, tubs, waterers) in shaded areas.

Arid

If you live here, you don't have to be told that water is your most important resource. Minimize evaporation loss and accidental spillage by establishing a monthly water-system maintenance check, testing soil-saturation levels along the waterline route and lowering floats on automatic waterers for lower flow. Horses will often play in water bowls that are too full. If they have to wait between large gulps, they will have less time to play and more water to drink.

FIELD AND PASTURE ROTATION

How to maintain pastures and protect the ecology with effective rotation and planting techniques

L AND IS A PRECIOUS COMMODITY, not just for growing crops but also for the health and well-being of your horse. Even if he's an expensive show horse, it is healthy for him to be outside, to get some sun on his back and some grass under his feet. Outdoor exercise in an area that allows freedom of movement has been proven to help with muscle development, digestion, hoof health, and skeletal health in ways that riding cannot.

So how can you best maintain your pastures and protect your unique ecology while keeping your land — and your horse — healthy? The answer is sound field and pasture management that balances the needs of your horse with the ability of the land to produce good-quality forage and pasture. Your horse will be healthier and your land more efficient at both production and recovery. You will reduce your feed bill as well — trucking in less hay not only means a smaller feed bill but also saves many pounds of carbon and other emissions from entering our atmosphere.

Land used for horse pastures is commonly neglected and often home to too many animals. Although you can pasture as many horses on a piece of land as you like (zoning considerations allowing), you won't be able to graze them there very long before grass runs out. On the other hand, a single horse can graze a pasture for a long time. So where's the balance?

YOUR LAND: ONE BIG JIGSAW PUZZLE

Chapter 3 discussed mapping your property using topographical and aerial maps. For our purposes in this chapter, a map that shows your property and fence lines so you can see where your paddock, pastures, and turnout pens are is sufficient. Draw your barn, and shade in road areas. It's like one big jigsaw puzzle, and you need to decide what goes where. This chapter will help you determine the ideal number of horses and the appropriate rotation for each area on your property.

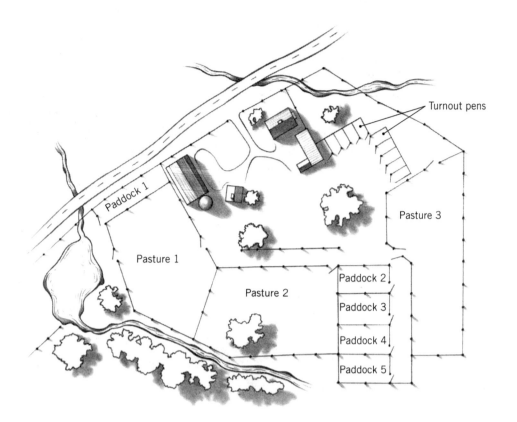

Turnout pens

Pasture 3

Paddock 1

Pasture 1

Pasture 2

Paddock 2
Paddock 3
Paddock 4
Paddock 5

*Whether your farm is large (above) or small (below),
a map will give you an overview of how efficiently
you are using it.*

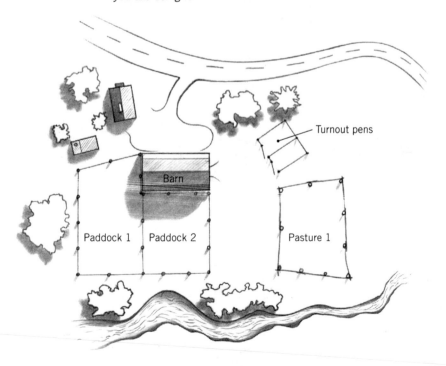

Turnout pens

Barn

Paddock 1 | Paddock 2

Pasture 1

Tally the number of horses you have on your property or how many you are planning to have. Mark this number in pencil so you can move horses around virtually.

Label each pasture, paddock, and field. Choose any method that helps you remember them. For the sake of clarity, we are going to use Pasture 1, Pasture 2, Paddock 1, Paddock 2, Turnout 1, and so on.

How Many Pastures Should You Have?

Wayne Burleson, a land-management consultant from Absarokee, Montana, says, "The short answer to this question is, more pasture is better." Flexibility is the key for Burleson, who travels around the world advising landowners how best to manage their pastures, taking into consideration their unique climate needs.

"Just think," he says, "if you have 36 pastures and each one will hold your entire herd for 10 days, that would equal year-round grazing with only 10 days of use on each pasture. I use a rule of thumb that you need at least eight pastures to get a decent rest period for fast growth-recovery time."

In his home state of Montana, the native prairie grasses have what he calls "30 honest growing days a year." During this time, you can maximize the time the grass is allowed to grow by not overgrazing it.

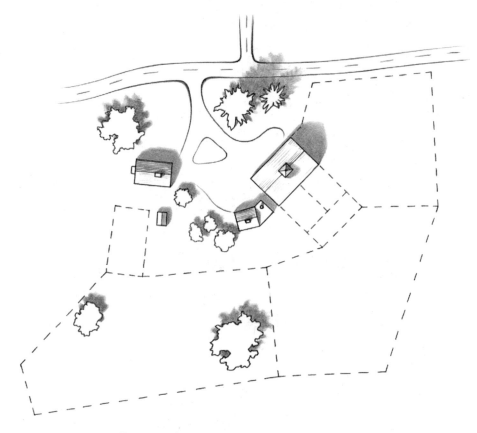

Proper grazing techniques are most effective when fence lines logically follow the natural contours of the land and the land has time to rest.

SUMMER PASTURE-MANAGEMENT SCHEDULE

For the majority of North American horses, summertime means increased pasture turnout. This also coincides with the peak growth time for your grass and forage. Alayne Blickle, director of Horses for Clean Water, has created a summer pasture schedule to help you manage your pastures most effectively.

MAY. Implement rotational grazing system. Never graze below 3 inches (7.6 cm), and don't allow grass plants to go to seed. Fertilize. Spread ¼ to ½ inch (0.6–1.3 cm) layer of compost or commercial fertilizer at rates recommended by your local conservation district. Mow weeds and tall grasses. Harrow manure in grazed pastures after rotating grazing areas.

JUNE. Rotate livestock on pastures. Graze to 3 or 4 inches (7.6 or 10.2 cm), remove animals, mow, harrow. Wait until pastures regrow to between 6 and 8 inches (15.2–20.3 cm) before regrazing. Mow and clip weeds and tall grasses. Harrow to spread manure after rotating livestock.

JULY. Rotate livestock. Inspect for weeds. Contact your conservation district office for help with identification and control methods. Continue harrowing to distribute manure and clipping tall grasses. Repair or build winter-confinement areas and covered manure-storage areas.

AUGUST. Conduct soil test. Take samples and send in for analysis (if not already done in spring). Continue to rotationally graze, mow, and harrow. Finish winter-confinement areas and covered manure- and compost-storage system.

SEPTEMBER. Fertilize. Spread compost ¼ to ½ inch (0.6 to 1.3 cm) thick or use commercial fertilizer, and spread at rate recommended by your conservation district. Finalize preparations for winter-confinement areas. Order any needed additional footing. Continue to rotate livestock and mow and harrow. Check gutters, downspouts, and outlets.

Burleson defines overgrazing as "the biting off of fresh regrowth, which is detrimental to healthy root development."

According to Burleson, the needed rest period in your particular area helps to determine how many pastures you should have. The drier the land where your pastures are located, the longer the recovery time needed to get back into a plentiful state.

Be careful if you have a large pasture with very few horses. You may think that you have light grazing going on, but hidden overgrazing happens. Some plants are overgrazed and don't get the chance to recover, while other, less palatable plants begin to grow out of control.

If you are concerned that doubling or tripling the number of pastures you have will require a lot of fencing, consider temporary fencing options. Solar electric fencing is an excellent, green option.

FLORA AND FAUNA

In case you don't remember from high school biology, flora refers to the plant life in an ecosystem and fauna refers to the animal life. For each enclosure on your property, you must determine what types of flora and fauna are present and if they are either helpful or harmful to the land.

Taking Inventory

When evaluating your pasture, you need to know what types of plants and grasses are supposed to be growing in your pasture and whether this differs from what is actually growing there. Books on local plant and grass species can be found at your library or gardening or farming stores. Once you identify the common native grasses and plants, list them on one side of a piece of paper and the unidentified or nonnative grasses and plants on the other.

If you need to take a sample, think back to those leaf projects your sixth-grade science teacher gave you: Press the leaf or plant sample between two sheets of waxed paper and stick inside an encyclopedia. This will preserve the sample until you can find out just what it is. You may be able to get help from a local university or even through an online search engine using descriptive terminology (for instance, "plant type, purple flower, pointy leaves" or another string of descriptors).

Knowing what's in your pasture is important for several reasons. First, you will be able to identify any dangerous plants that may harm your horse. Second, you will know what types of weeds may be around. Third, you can determine what sort of nutritional value your pasture grass has so you know if you need to supplement your horses.

One of the best ways to accomplish this final point is to have a pasture and forage analysis done by a lab that specializes in forage analysis (your local Extension office or university will know where you can find one). Walk diagonally from one corner of the pasture to the other and stop every 10 feet (3 m), take some clippings (with scissors), and move on another 10 feet. You'll end up with a big bag of grass that you can immediately stick in the freezer to halt plant respiration and prevent any chemical changes as the grass wilts. Then ship it directly to a lab as quickly as possible.

If at all possible, arrange to speak with a nutritionist to help you interpret the results. You'll see information on the specific nutrients in your pasture as well as the protein, fiber, starch, and sugar levels. A nutritionist will be able to tell you what nutrients and minerals might be missing so you can supplement accordingly.

Managing Weeds

Most horses will simply ignore weeds in the pasture because they are not as palatable as the grasses. In fact, that's one of the weed's defense mechanisms. When you're a plant, the only way to survive is to avoid being eaten, get enough water and nutrients, and keep growing. When horses ignore distasteful weeds, they actually create a better environment for those weeds to survive because eating the surrounding grass means less competition for nutrients. So what can you do?

Herbicides are usually waterborne chemicals that target specific plants and then degrade quickly so that other plants can grow. They are often applied with mechanical spreaders. However, herbicides can be applied improperly — just before or after a rainfall — and can spread into the water supply.

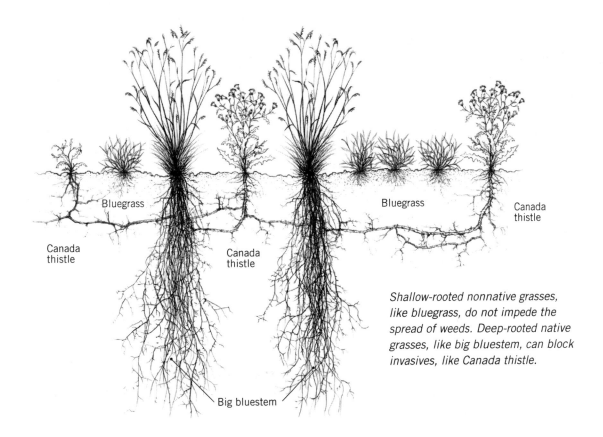

Bluegrass

Canada
thistle

Canada
thistle

Canada
thistle

Bluegrass

Canada
thistle

Big bluestem

*Shallow-rooted nonnative grasses,
like bluegrass, do not impede the
spread of weeds. Deep-rooted native
grasses, like big bluestem, can block
invasives, like Canada thistle.*

When looking for alternatives to chemical herbicides, keep in mind that almost any herbicide can be called "organic" because carbon is the primary molecular component (except in arsenic-based herbicides). Genuine organic herbicides cost more money and are often less effective based on application alone. Sometimes secondary measures such as hand weeding are necessary. Most organic weed-control options have been developed for use by organic vegetable-crop farmers because of their commitment to pesticide- and herbicide-free produce. These herbicides can all be adapted for larger pastures.

Six Alternatives to Herbicides

It's important to note, however, that not all organic herbicides are safe: organic does not necessarily mean "nontoxic." Many organic herbicides contain ingredients that damage plants and animals alike. Following are some nontoxic options.

The Most Effective Weed Control

You'll notice we didn't say "most efficient" form of weed control because the most effective method takes muscle power: cultivation. That's the mechanical or manual removal of weeds from the root. This can be done with a cultivator (though that will destroy the other plants in the area as well) or with a shovel, one weed at a time.

Map your weeds so you can make sure they are being eradicated and to help you notice any weed patterns. Weed distribution is not uniform; patches of weeds, particularly perennials, should be managed individually. Mapping the weed patches makes follow-up evaluations easier, so additional treatments can be applied as needed.

1. VINEGAR

Not just for potato chips and French fries any more, according to the Agricultural Research Service of the United States Department of Agriculture (USDA), vinegar is an effective weed killer for organic farmers:

> (Researchers) hand-sprayed the weeds with various solutions of vinegar, uniformly coating the leaves. They found that 5- and 10-percent concentrations killed the weeds during their first two weeks of life. Older plants required higher concentrations of vinegar to kill them. At the higher concentrations, vinegar had an 85- to 100-percent kill rate at all growth stages. A bottle of household white vinegar is about a 5-percent concentration.

Canada thistle, one of the most tenacious weeds in the world, proved the most susceptible; the 5-percent concentration killed 100 percent of the perennial's top growth. The 20-percent concentration can do this in about 2 hours.

Spot spraying of cornfields with 20 percent vinegar killed 80 to 100 percent of weeds without harming the corn, but the scientists stress the need for more research. If the vinegar were sprayed over an entire field, it would cost about $65 per acre (0.4 hectare). If applied to local weed infestations only, such as may occur in the crop row after cultivation, it may only cost about $20 to $30.

The researchers use only vinegar made from fruits or grains, to conform to organic farming standards.

2. WALNUTS

As we learned back in chapter 3, walnuts are a natural herbicide. More specifically, the natural compound juglone is toxic to many other plants. It's an allelopathic compound, which means it is synthesized by one type of plant and affects the growth of another. Alfalfa and crimson clover are sensitive to juglone, while white clover, red top grass, orchard grass, soybean, timothy, and wheat are not. There is continuing research on walnut-based herbicides and their effectiveness in controlling weeds. Often walnut hulls are used for weed control but are not generally used on pastureland because of their toxicity to other plants. Oil from the leaves of black walnut trees is sometimes used in the production of herbicides.

3. STEAM

Using steam to kill weeds may not be effective on all types and may require repeated use by trained professionals. Powered by diesel, the steam is applied at more than 350°F (177°C).

4. FIRE OR THERMAL

Applied in short, controlled bursts and powered by propane, thermal weed-control devices can be very effective, especially when dealing with tough weeds, such as stinging nettle and poison hemlock. There is risk, however, to surrounding areas and wildlife — this is fire, after all — so this method should never be used during dry seasons.

5. INFRARED

According to the National Sustainable Agriculture Information Service:

> Infrared weeders, first developed in Europe, are heated by a propane torch, but the flame is directed toward a ceramic element or steel plate that radiates at temperatures of 1800 to 2000°F [982–1090°C]. The danger associated with an open flame is thereby minimized. The mechanism of weed control is the same as in flame weeding; cell contents — plasma and proteins — are disrupted and the plant wilts down and dies. Infrared heaters are available in handheld, push-wheeled, and tractor-mounted models.

In addition to weeding, the tractor-mounted infrared thermal units are used to control Colorado potato beetle and potato vine desiccation. Some of the tractor models feature the injection of forced air to increase the effect.

6. MULTISPECIES GRAZING

Use of one or two goats on your property can help control weeds, but you should ensure that the goats are properly contained, and you may need to adjust your fencing to do so. Some goat producers actually market their services to landowners with a large number of acres — the producer will provide the goats and herders (including dogs or horses) to manage weed growth at critical times each year.

Goats are effective because once they graze the weed it is unable to go to seed without the flower on it, and it cannot photosynthesize to build a root system without leaves. Goats like such weeds as Canada thistle, cheatgrass, common candy, common mullein, Dalmatian toadflax, dandelions, downy brome, Indian

Hire a Goat

Utah State University experimented with using goats to reduce noxious weeds in large areas and found that

> results indicate that grazing at the bud-to-bloom stage has the greatest potential as a control tool. Grazing at the rosette-to-bolt stage does reduce seed count, plant count and canopy cover, but not at the levels of bud to bloom. Grazing twice reduces seed heads the most but results in increased plant count, perhaps because grazing disturbs the seed bank causing quicker germination or because the goats don't eat the dry seed heads, instead knocking them to the ground.

tobacco, knapweeds, larkspur, leafy spurge, locoweed, musk thistle, oxeye daisy, plumeless thistle, poison hemlock, purple loosestrife, Scottish thistle, snapweed, sweet clover, yellow star thistle, and yucca.

Goats and horses make good companions both socially and ecologically.

Creating a Healthy Habitat

After all is said and done, however, you may not be able to avoid the effects of chemicals if your neighbors use them. Your best bet is to create natural barriers around your property using shrubs and natural and man-made land formations and diversions.

 Homemade Herbicide

Trying to beat those weeds? Here's a homemade herbicide recipe for tank sprayers or spray bottles that's low in chemicals and caustic substances.

1 gallon (4 L) household white vinegar
1 tablespoon (15 mL) Turbo Spreader Sticker Concentrate — purchase at garden or home centers (If you want strictly nontoxic weed control, omit this item from the recipe.)
1 ounce (30 mL) insecticidal soap concentrate — purchase at garden or home centers

Mix ingredients together in a tank sprayer, and spot-spray weeds on a calm day when it is not going to rain for at least 12 hours.

Note: The spreader sticker is a useful product even for those who use chemical herbicides. When used as an additive to sprays, it acts as an adhering agent and helps disperse product evenly, protecting it from rain and sun and keeping it on the plants, so it doesn't run off or evaporate. Also, you won't have to use as much insecticide/herbicide to get the job done!

Source: HorsesForCleanWater.com
(used by permission)

BARN SWALLOWS

Barn swallows, the most common insect predator, can consume a large number of insects, especially during the spring. To attract them to your barn (or outbuilding, which usually offers less nest disturbance), buy a premade swallows' nest to install under overhangs. The Maryland Cooperative Extension at the University of Maryland offers a handy guide to creating your own barn swallow nest (see Resources).

A family of barn swallows will drastically reduce your farm's population of flying insects.

BATS

Another awesome and hardworking insect predator is the bat. If you're afraid of bats or consider them disease-ridden, check out what Bat Conservation International has to say:

Bat rabies accounts for approximately one human death per year in the United States. Thus, some people consider bats to be dangerous. Nevertheless, dogs, which often are considered "man's best friend," attack and kill more humans annually than die from bat rabies in a decade. Statistically speaking, pets, playground equipment, and sports are far more dangerous than bats.

Clearly, bats do not rank very high among mortality threats to humans. Nevertheless, prudence and simple precautions can save lives.

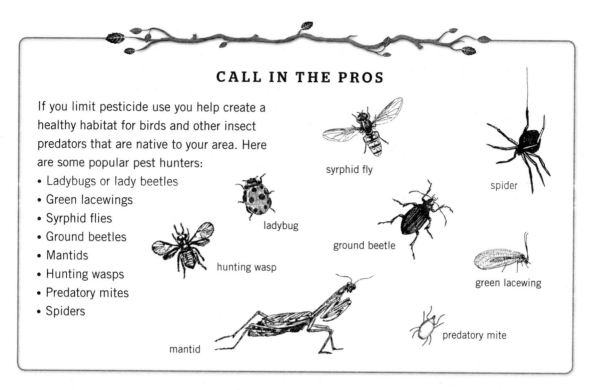

CALL IN THE PROS

If you limit pesticide use you help create a healthy habitat for birds and other insect predators that are native to your area. Here are some popular pest hunters:

- Ladybugs or lady beetles
- Green lacewings
- Syrphid flies
- Ground beetles
- Mantids
- Hunting wasps
- Predatory mites
- Spiders

syrphid fly

spider

ladybug

ground beetle

green lacewing

hunting wasp

mantid

predatory mite

Some bat species consume up to 1,000 mosquitoes an hour.

Consider setting up a bat house on your farm, not only to help with bat conservation but also to reduce the number of insects on your property. Bat Conservation International has excellent resources online for locating, purchasing and installing a bat house (see Resources).

GARLIC

Garlic has also been fed to horses to deter flies and other insects. The strong-smelling substance is either sprayed upon the horse or ingested. After eating the garlic the mildly noxious odor is excreted through the pores in the horse's skin and dissuades the insects from landing or biting.

SOIL AND FORAGE HEALTH

As you learned in chapter 3, multiple soil types are common on every property. Some types, along with soil health, will help resist erosion and maintain good pasture quality. If you have tested your pastures, note the types of soil there. Soils with a higher clay concentration are more resistant to erosion than soil with high sand or silt content.

But more important than the type of soils or grasses you have is how healthy they are. Overgrazing is the number-one problem that contributes to unhealthy pastures, and soil health is right behind it as a factor (see chapter 3).

CREATING A SACRIFICE AREA

A sacrifice area — a paddock or pen used to house horses when they need to be kept off pastures — is a simple yet vital part of pasture management. It's called a sacrifice area because you sacrifice it for the good of your other pastures: to protect them from damage due to grazing on wet, muddy ground or overgrazing during the summer. It can also be used to control how much grass or feed a horse has access to. Sometimes it is referred to as the "diet pen"; horses in the sacrifice area cannot freely feed on grass because there isn't any, especially if you choose to put down footing such as chips, gravel, or hog fuel.

Of course, you must ensure that the area is safe and set up much like a regular paddock.

Some important attributes of sacrifice areas are the following:

- At least an acre per horse
- Safe, secure fencing
- Safe places to feed that take the footing into consideration (for example, you can't feed hay on sandy footing as it may result in sand colic)
- An adequate water supply
- Enough shelter for the number of horses housed

HOW TO USE A SACRIFICE AREA

The sacrifice area can remain empty when the weather is fine and pastures are healthy, when overgrazing is not an issue. But during the rainy season, a saturated and rain-soaked field can be damaged in just a few hours. Horses' hooves break the roots of the grass and compress the soil, suffocating any roots not already destroyed. Hooves also disturb the topsoil, mixing it with the rainwater and creating mud. When this mud dries, it becomes dust, and valuable topsoil simply blows away.

During periods of drought you may also use the sacrifice area to allow time for your pasture to rest and grow. Overgrazing can become a chronic issue if left unchecked. Healthy grass is eaten down, and weeds are left to flourish.

LOCATION, SIZE, AND LAYOUT

Establish your sacrifice area on high ground located in a convenient area for doing chores. You won't be able to rely on the grass inside the enclosure to provide nutrients, and you'll have to haul feed to your horse when he's there. Also, you must clean the area regularly (at least weekly) by picking up the manure. Because this area will likely be much smaller than the larger pasture or paddock that you are relieving, manure and urine can pile up quickly. Leaving manure on the ground will increase your horse's parasite load and attract flies. During the rainy

Temporary fencing can turn a corner of your paddock or pasture into a sacrifice area during times of year when you need to protect your fields from hooves and grazing.

season, manure left lying on the ground can be washed away and may contaminate surrounding fields or water sources.

The ideal size of the sacrifice area will vary greatly depending on the number of horses you have. If you have fewer than five, it's possible that horses can share one smaller area, or you can establish a series of smaller pens with one horse per pen (probably the safest plan, unless your herdmates are all buddies).

It may be easiest to cordon off the corner of the pasture that is home to the water source, whether it is an automatic waterer or a large tank. This area sees the most foot traffic anyway and is likely to have less grass no matter what you do.

P - A - S - T - U - R - E

Wayne Burleson of Absarokee, Montana, has been offering consultation services to landowners all across North America for more than three decades. He is a professional land-

management consultant certified by the Society for Range Management and has a master's degree in agriculture, range science. Utilizing his unique, holistic approach to land management, he has developed seven key principles to a healthier pasture.

P = DETERMINE THE AREA'S PURPOSE.

Know the main purpose of the area to be grazed. Examples: profit from livestock grazing, attract wildlife, use as shelter, enjoy its aesthetics, or special use. Why is this important? If you have undetermined purpose, your management decisions will be without direction. Don't skip this step!

A = THE AMOUNT OF TIME SPENT GRAZING MUST BE CONTROLLED.

Controlling the amount of time spent grazing in each unit (the frequency of grazing) in different seasons (the seasonality) is just as important as, or perhaps even more important than, what the animals are eating (their selectivity) or the amount of forage removed (the intensity-utilization levels). You need to balance all of these key factors to have healthy pastures.

S = SUPPLY ADEQUATE REST TO KEY PLANTS FOR FULL RECOVERY.

If you allow continuous grazing to occur on fresh regrowth on certain high-producing plants, their roots will start to shrivel up and the plants will begin to die. Make sure to supply adequate rest to key plants after each bite. Monitor their recovery period. How? Place a flag by a previously grazed plant and monitor the number of days for full recovery. Consider more than one year's rest to bring back needed plants to full vigor. Also, try "flash grazing" at low stock density, rotating from pasture to pasture as fast as possible; use this grazing management on very early spring pastures.

T = TEST NEW IDEAS.

Experiment with small test sites before implementing major changes in larger areas. Each pasture responds differently to different grazing strategies. Construct enclosures (around areas with no grazing) to evaluate the differences in grazing treatments. This will let you compare grazed areas to those not grazed. Example: Test "herd effect," the trampling down of old vegetation to feed the soil, and then monitor the results. If you find good results in smaller areas, implement changes in larger ones.

U = USE LITTER TO FEED THE SOILS.

Lack of available water is the biggest limiting factor for most plant growth. We cannot control the amount of rain received, but we can control grazing. Use the following rule of thumb: Keep the soils covered. Plan pasture management to feed and shade the soil. Litter such as dead leaves that fall naturally to the ground, compost or remnants of hay feedings will lower soil temperatures to conserve water. The old grazing rule of "taking half and leaving half" works on the volume of forage in the drier, low-producing areas. However, this rule may waste forage in high-producing pastures. Plan land-reclamation work during the non-plant-growth season. Winter grazing is a great time to feed the soils.

R = RELY ON DIVERSITY.

To increase diversity, use several different grazing methods, systems, and strategies; strive

The Proper Setup

By now you may be asking, "What does the proper setup look like?" There are as many answers to that question as there are acreages, ranches, and farms in North America. Each farm will have a unique design that can be maximized for efficient use.

You will know that the setup is right for your farm if:

- The horses are healthy and happy to come inside at night and go out to the fields in the day (they are receiving adequate grazing time)
- You are able to maintain good forage and grass growth year-round (unless it's covered by snow)
- Wet paddocks do not turn into muddy paddocks
- Weeds are the exception, not the norm
- Extreme weather does not cause extensive damage to your pastures and paddocks

If the above statements describe your setup, congratulations. You have the right resources to manage your paddocks and pastures, and you are able to work with your horse's environment to the benefit of both the land and the animals.

to increase the number of different plant species and fine-tune your pasture management year after year. Why use diversity? One example is that warm-season grasses provide green forage later in the season than do cool-season grasses. Learn your area's key indicator-plant species. This will help you recognize when important changes are occurring. Install photo points. Locate and mark areas on the ground so that you can keep coming back each year to study the changes.

E = EVALUATE RESULTS.

Monitor, monitor, monitor. Use the feedback-loop method of monitoring: plan, control, monitor, and replan. Fine-tune each area by walking pastures often, looking for any signs of stress occurring on plants. Mark these problem areas. Make needed changes in management to fix problems early. Consider carrying over old forage for next year's early-spring-use pastures.

Don't graze certain areas during the second rotation. This will increase the energy flow back into the soils. Add a pasture reserve for drought emergencies. Determine the root causes of the problems and fix them; don't treat the symptoms. The more pastures you have, the more flexibility you'll have in creating healthier pastures.

☼ ❄ CLIMATE VARIATIONS

Subarctic

With the shortened growing season in the subarctic climate zone, plants and forage need that critical window of growth in the spring so they can get a foothold and establish themselves for the summer. Pulling your horse off pastureland for the early part of spring when the snow is melting will go a long way toward protecting root systems and soil retention.

Humid Continental

In the humid continental region, having one or more sacrifice areas will help your property stay healthy. Fluctuation in weather patterns can produce more extreme weather and shorten the recovery time for pastures.

Humid Oceanic

With the narrow range of temperature changes across the seasons in the humid oceanic zone, it can be tempting to overgraze because you do not notice the gradual depletion of available forage. Carefully monitoring forage height and sticking to regular rest periods will help your pasture remain healthy.

Highlands

Snowmelts can be sudden and drastic in the highlands region, so make sure you have your sacrifice area ready as early as March, in case a Chinook blows in and the snow melts within a few hours.

Semiarid

The heat of the day in the semiarid climate zone means your horses will seek shelter more frequently. Because of this, the area around shelters or loafing barns can become taxed from increased hoof traffic and increased manure. If possible, set up multiple shade areas for your horses using constructed shelters, lean-tos and tree stands, and make sure you pick up the manure regularly.

Arid

It's very easy to overgraze pasture in the arid region. Horses might benefit from being turned out at night (when they will eat and move around less) and being brought in during the day (as long as the barn is cool).

ECOFRIENDLY FEEDING

Feeding 1 to 100 horses while minimizing

your carbon footprint

As we've discussed throughout this book, incorporating environmental awareness into your horsekeeping strategies improves the world and helps your horses lead healthier lives. The impact is seen on your property with healthier pastures and animals but also has wide-ranging effects at the regional, national, and, yes, global level. Every little bit helps. Nowhere is this more apparent than in your pastures and fields, where you can actually see the "green" effect.

Many counties throughout North America restrict the number of horses allowed per acre. Allowances are based sometimes on pasture acreage and sometimes on the acreage of the whole property. Some zoning allows for concentrated operations, where horses spend half their day inside a stall. Each county's regulations are different, so call and ask what they are in your area.

A standard rule of thumb is one acre per horse. If you have stalls and daily turnout and are utilizing your pastures effectively, you can house more animals, but as you've learned in this book, you must manage manure and pasture rotation efficiently.

In all parts of the continent, it becomes necessary to feed some type of roughage such as hay sometime during the year. Not many horses eat only grass all year round. Not only is the nutrition in grass alone inadequate, but few of us have enough land to pasture horses all year long without feed supplementation.

GROWING YOUR OWN FEED

Growing your own feed gives you the greatest control over the carbon outputs used in seeding, harvesting, and baling. If you do not have enough land, consider establishing a co-op with other horse owners in your area. If one person has enough land and five property owners can pool resources (equipment, labor, transportation, storage), you will also be able to minimize your impact.

Make sure if you are growing your own hay and forage that you practice low- or no-till farming (see the Definitions box opposite),

If you have enough land to grow your own hay, pool resources with your neighbors at haying and other labor-intensive times, as was traditionally done since farming began. This will minimize your impact on the environment as it strengthens your community.

conserve fuel, and use minimal fertilizers or nothing other than compost.

If you are considering becoming an organic farmer, you will need to investigate the standards of practice for organic farmers. In both the United States and Canada there are specific guidelines that you must follow to be considered a true organic farmer, including a very long "permitted-substance list."

Being "green" is not synonymous with being "organic" because organic has specific guidelines for how the food was grown and what was used to grow it and a host of other practices and standards (see box). North America has only 7.2 percent organically grown cropland. Europe, on the other hand, has 23 percent and South America 19 percent.

Alternatively, you can become "certified naturally grown," which with fewer regulations, is more appropriate for small-scale organic farmers. Find out more at the Organic Ag Centre or the Agricultural Marketing Center (see Resources). But being greener doesn't need any kind of certification: it can start with changing just one behavior at a time.

Organic Practices

Some examples of regulated practices are:

- Composting methodology
- Pesticide use
- Water management including runoff
- Protection of crops from "genetically modified organisms" or GMOs
- Healthy soil maintenance and conservation
- Prevention of nutrient leaching

DEFINITION: No-till

Also known as direct seeding, no-till is defined as planting directly into the residue of the previous crop, without tillage that mixes or stirs soil prior to planting. No-till reduces soil erosion, improves soil structure and organic matter, and reduces fuel inputs.

If your place has adequate space for setup and parking, it might be ideal for a regular or occasional farmers' market.

If it isn't feasible for you to grow your own food, consider supporting local farmers. If you have some spare land, think about hosting a semiannual farmers' market. You may be able to get free produce for your family while providing an opportunity for those around you to buy locally as well.

EVEN HORSES CAN EAT LOCALLY

In April 2007 Alisa Smith and James MacKinnon published *The 100-Mile Diet,* a book that explored the possibilities of eating only food that was born or grown within 100 miles of your home. Canadian authors Smith and MacKinnon learned that the food they bought at the local supermarket usually had to travel the distance between Toronto and the Yukon (3,500 miles or 5,500 kilometers) before it reached their table. The fuel costs for each meal counterbalanced all the good they were doing trying to be conscious of the environment and live a green lifestyle.

Can we provide local feed for our horses? For those living in farming country, the answer is very likely yes. According to the United States Environmental Protection Agency, more than 60 million acres of hay is grown each year in the United States, the majority in the Midwest. And in Canada there is almost 70 million acres of forage crops — and about 20 percent is exported to the United States annually. The majority of these "breadbaskets" are located in the middle of the continent. So what are horsekeepers in other regions to do if the hay is so far away?

Planning for the future begins with your small farm. As overdevelopment spreads from the cities out to the rural areas, farms are disappearing. At the same time, horses are actually moving closer to the cities because now they are used more for recreation than farming. This creates an inverse relationship: farms (and thus hay) moving farther from the city, horses moving closer.

The Communal Hay Shed

We need to support the farmers closest to us that are still producing hay so that producing this crop remains profitable. As a community, you and your neighboring horse owners can plan to build larger hay sheds with good road access so that farmers can deliver entire tractor loads to one location. This means they spend less money on fuel and less time on the road, and horse owners (who would otherwise have smaller hay sheds and more hay deliveries) will spend less in fuel charges to receive the hay.

Our farm-building habits have taught us to build hay sheds to suit the individual property, so we have multiple small hay sheds on many small farms. Consequently, some smaller farms are finding it difficult to locate farmers who will deliver multiple loads of hay because the fuel costs are simply too high. Joining with other horse owners in ordering a larger load of hay to one location is a good way to make sure everyone wins.

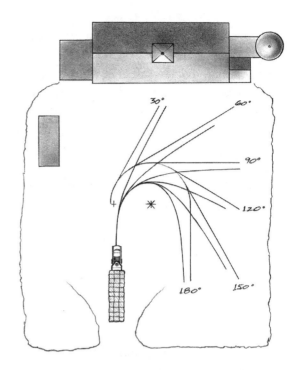

In planning the site, design, and dimensions of a hay shed, be sure to allow adequate space in the driveway for a large delivery truck to turn around and unload.

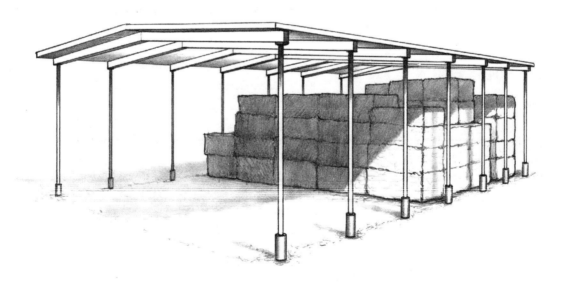

A larger hay shed can be used for communal storage for surrounding horse properties; however, no equipment should share space with hay unless the proper footing (cement pad, cordoned off from hay) can prevent contamination from vehicles.

Alex Aballo, project director for Equestrian Services LLC, has begun to notice that farmers and ranchers are cooperating on their hay storage on an informal basis: One person warehouses the hay for the rest of the people in the community. Perhaps the others pay rent, or perhaps the person warehousing the hay takes a cut from the rest. A neighbor stores his hay in Aballo's hay shed, which is large, located right off the road, and has a big turnaround in front. She reports, "My neighbor has already found that he can't get small deliveries because it's just not worth it for the farmers any more; they're too far, and they're spending too much gas just trying to get to us."

Equestrian Services LLC recently began building an equestrian facility in Lake City, Florida, a specialized community that includes homes and trails surrounding a riding facility. Because the "farmettes" in this equestrian community are no larger than five to ten acres and cannot have a sizable hay shed on their properties, the company is building a central hay facility that will accommodate a load of hay large enough for all.

GREENHOUSE GASES AND YOUR FARM

As mentioned in chapter 2, greenhouses gases (GHGs) are important to the survival of the earth because they allow warmth to be trapped inside our atmosphere. Because atmospheric concentration is increasing, however, and raising the temperature of the earth, scientists have begun to notice the phenomenon of "climate change": a change in climates that have remained relatively stable since the last ice age, 10,000 years ago (give or take a decade).

The principal greenhouse gases are nitrous oxide, methane, and carbon dioxide, and the last, carbon dioxide (CO_2), is the main one emitted by industry. While cattle ranchers have cornered the market on methane emissions, horse owners also contribute GHGs, most prominently nitrous oxide from manure handling and storage and from commercial fertilizer. A breakdown of GHG emissions by Agriculture and Agri-Food Canada in 2001 showed that 24 percent of emissions came from fertilizer use and 12 percent from crop management. (See more on this subject in chapter 2.)

Greenhouse Gases and the Agriculture Industry

Because growing crops and livestock for food is deemed a high priority and heavier regulations would put a financial strain on many producers, governments have done little to regulate the agriculture industry in regard to pollution. Since 10 percent of all greenhouse-gas emissions in Canada come from the agricultural sector, however, the Greenhouse Gas Mitigation Program for Canadian Agriculture recently received $21 million "to address agricultural GHG emission reduction and carbon sequestration enhancement in the areas of soil, nutrient and livestock management."

Within the group of producers of food crops and cattle, the focus has been on cattle methane emissions, the number-one source of greenhouse gases in the agricultural sector. Much of

---

How Farms Contribute GHGs

The agricultural industry contributes its share of greenhouse gases. Methane is the largest, mostly from the cattle industry. Carbon dioxide is the next-largest emission.

In the horse industry the two biggest emitters of carbon dioxide are:

1. Transportation: fossil-fuel combustion
2. Utility use: electricity

the research and applications, such as the following, can be applied to horse owners as well:

- Grazing management strategies can increase the quantity and quality of forages on pastures and native rangelands
- An increase in the quantity of forage can increase the amount of carbon sequestered in soils
- Feeding management strategies can increase the efficiency of feed utilization
- Manure management strategies can preserve the nutrient content of manure

Six Strategies to Help Restore the Carbon Balance

In the previous chapters we've discussed how to diminish our dependence on fossil fuels and reduce our energy costs. Now we'll focus on ways to sequester carbon; in other words, to create "carbon sinks" that help capture carbon to lock it out of the atmosphere.

What can you do to decrease carbon dioxide output and increase carbon sequestration? Here are a few ideas:

1. **Grow your own hay and grain** if you can. This reduces the number of miles that feed has to travel to get to your barn door.
2. **Buy hay from farmers that use low- or no-till farming practices.**
3. **Implement a grazing plan** for your property, and do not overgraze. Healthy forage can increase carbon sequestration.
4. **Develop a plan to keep horses away from sensitive riparian areas.** In addition to protecting water quality, this plan will protect the highly productive vegetation on riverbanks and lake edges that is capable of sequestering more carbon than it emits.
5. **Use proper and timely manure-composting techniques** instead of fertilizing (see chapter 7) to increase the production of pasture grasses. This can sequester more carbon and increase the number of grazing days. To prevent runoff and increase efficient use of nutrients, pay attention to the proper conditions for composting manure.
6. **Plant shelterbelt trees,** which can remove significant amounts of carbon dioxide from the atmosphere and store it as carbon in the wood and roots over the years.

Trees grown in windbreaks and shelterbelts offer many benefits to your property: they provide shelter to animals, reduce wind flow, limit soil erosion, control snowdrifts, produce fruits and nuts, create wildlife habitats, and sequester carbon.

Carbon dioxide: The most common greenhouse gas. It is produced during the aerobic (in the presence of oxygen) decomposition of organic matter and plant and animal respiration, as well as through the combustion of materials and fuels. This gas is removed from the atmosphere through photosynthesis and ocean absorption.

Carbon sequestration: The capture and storage of carbon dioxide in plant material or the ocean. As part of photosynthesis, plants remove carbon dioxide from the air, strip the carbon out of it, and produce leaves, branches, and roots. Plant residue (carbon) is returned to the soil, thereby increasing soil carbon.

Carbon sink: A place where carbon accumulates and is stored. For example, plants are carbon sinks. They accumulate carbon dioxide during the process of photosynthesis and store it in their tissues as carbohydrates and other organic compounds.

LOOK AT THE ENTIRE PICTURE

Just as with many other green behaviors in this book, you need to look holistically at what you are doing on your property or with your horse to see how it affects the cycle as a whole. As Bobbi Helgason, a biochemist with Agriculture and Agri-Food Canada, noted in a 2004 Alberta Environmentally Sustainable Agriculture forum, "Rather than looking at emissions from one source at a time, we want to follow nutrient flows across the whole farm so we have a better idea about how changing practices might affect net emissions across the entire farm landscape."

She adds, "One example I often use is the planting of alfalfa. We know that planting perennial legumes can increase carbon storage in our soils. However, we need to keep in mind that those legumes will be fed to cattle, which will increase production of enteric methane (in the digestive system of cattle). Some of those nutrients will be excreted in manure, and how we handle that manure will also affect greenhouse gas emissions."

CARBON CREDITS Q&A

To combat the worldwide problem of increasing greenhouse gas emissions, a credit-based system was created to reward companies that reduced their output of GHGs and to increase the number of projects that would remove GHGs from the atmosphere. A system of assigning "carbon credits" meant that if a country (or company) had a large carbon footprint they could not reduce effectively, they could purchase a "credit" from a company that had earned the credits by reducing their own emissions or creating carbon sink projects.

Q: *What Are Carbon Credits?*

A: Every business must take stock of inputs and outputs to survive: They seek inputs (payments from customers), and they are responsible for outputs (their product or service). Carbon is produced in every major industry, and the carbon-credit system moves carbon into the "output" section of business.

Companies and industries that have the highest amount of greenhouse gas (GHG) emissions may eventually have limits imposed under future international agreements, such as the Kyoto Protocol. Currently, they can choose to reduce their output to meet those limits or choose to purchase carbon credits in the amount of their shortfall.

Q: *Who Buys Carbon Credits?*

A: Businesses that have extra credits can sell them to businesses that need to purchase them, or they can trade them on the international market. There are businesses and organizations that are set up specifically to sell carbon credits.

These businesses, called carbon projects, must be approved through such initiatives as Clean Development Mechanism (CDM) or Joint Implementation (JI) and must provide activity reports to prove they are actually reducing greenhouse gases. Projects may include reforestation, carbon capture and storage, or wind and solar projects.

Q: How Much Is a Carbon Credit?

A: A credit represents one metric ton (1.1 tons) of CO_2 emissions. There are four international exchanges where these can be bought and sold: the Chicago Climate Exchange, the European Climate Exchange, NordPool, and PowerNext.

Q: Who Sells Carbon Credits?

A: The sellers of carbon credits are companies or producers who are reducing their GHGs and can sell their accrued credits through a brokerage. Says Rudy Saufert of CGF Brokerage, Saskatoon, Saskatchewan, one way to sequester carbon is with farming practices that collect carbon out of the air and put it into the ground.

One of those practices, he says, is no-till, zero till, or minimum tilling. Minimum tilling is defined as using 3-inch (7.6 cm) openers on 9.5-inch (24 cm) spacing while tilling. This practice ensures that carbon is not being released back into the air.

In Canada two kinds of credits are available: credits for reducing emissions and credits for removing carbon dioxide from the atmosphere (as when reducing tillage).

For example, emissions reductions must be:
- **Additional.** Your emissions are lower than before you changed your practice.
- **Real.** They result from a demonstrable action.
- **Measurable.** They can be measured with accepted methods and verified by a third party.

- **Clearly owned.** The seller must have clear ownership rights.
- **Permanent.** They will not be reemitted later, or if they are, there's a contingency plan in place.

Carbon Credits and You

So do you have carbon credits that you can sell? If you have read this book cover to cover and found ways to reduce carbon emissions and remove CO_2 from the atmosphere, you may be able to apply to sell these carbon credits, as long as the above criteria have been met. The Web site of the Chicago Climate Exchange (www.chicagoclimatex.com) lists the types of projects that may qualify:
- Agricultural soil carbon offsets (conservation tillage, grass planting)
- Energy efficiency or fuel switching
- Forestry carbon (afforestation and managed forest projects)
- Renewable energy systems, which offset fossil fuel–based electricity
- Rangeland soil carbon offsets (restoration or management of rangeland over minimum five-year commitment in designated regions)

☼ ❄ CLIMATE VARIATIONS

Some areas of the country will naturally emit more greenhouse gases because they need to use more energy to either heat or cool their homes. Those who live in the extreme north or south of North America cannot survive without proper heating or cooling mechanisms. Also, if your property is located in a secluded rural area you may already have established strategies to minimize your travel into urban centers. Many who live close to a city, on the other hand, think nothing of a few trips into town a week. When that becomes a few trips a month, you have truly become more efficient.

Subarctic

The short growing season in the subarctic zone may hinder abundant, healthy crop growth; instead you may have to import hay from far away. If you are located near a railway, join with other landowners to arrange for a large hay load to be shipped by rail.

Humid Continental

You are located in the humid continental region, the breadbasket of the world, where most of the continent's hay and food are produced. It's much easier for you to buy locally, but keep your ear to the ground to find out which farmers are considering switching crops to biofuels, and support their efforts to grow hay instead. Even if you have to pay a little more, it's worth it to keep hay growers in business.

Humid Oceanic

In the humid oceanic climate zone, you may be able to get some good-quality second- or third-cut hay near the end of the growing season. While you may think this would result in lower hay quality, nitrogen toxicity will be less likely, as most fertilizers are applied just once rather than between cuts.

Highlands

In the highlands most of your hay will be grown along the region's borders in the foothills, so it may be difficult to buy locally.

Semiarid

This zone is often affected by droughts. Ensure that you have backup plans for forage; use cubes as supplements, and have some on hand throughout the summer.

Arid

If you live in the arid climate zone, you need to monitor haystacks for moisture content. Any stack of square bales with a moisture content of 16 percent or greater is in danger of catching fire. Check the temperature of the hay daily if it's above 120°F (49°C); twice a day if it's 140 to 150°F (60–65°C), say the experts. At 150°F the hay is entering the danger zone; check the temperature every two hours. If it's between 150 and 160°F, start moving hay out of the stack. At 160°F (70°C) or higher, call the fire department.

THE CHEMICALLY CHALLENGED HORSE

An overview of the common chemicals we use on our horses

with alternatives and disposal techniques

YOUR HORSE'S WORLD typically consists of paddock, pasture, barn, and arena. In each of those areas, chemicals continually challenge him. Most of the time we use chemicals — dewormers, medication, pesticides, fertilizers — with good intentions, but they can produce undesirable results. We may not even be aware of where the chemicals come from and what damage they might be doing. (For tips on creating a healthy habitat and using fewer chemicals on your own land, see chapter 9.)

Effects on Horses

What are these toxins doing to our horses? To find the answer, I consulted Mary Ann Simonds, BS, MA, who has worked professionally in the horse industry for more than 30 years, riding, training, and teaching. Currently, she directs the Equestrian Science Institute and the Institute for Integrated Sciences, in Vancouver, Washington, offering coaching and consulting in human-animal consciousness, ethology, ecology, and psychology.

Detoxing

Mary Ann Simonds (a wildlife and range ecologist and equine behaviorist) recommends 30 to 60 days of detoxification if you believe your horse may have consumed toxic hay. She prefers to give horses herbs that specifically target the liver and kidneys as well as homeopathic remedies and advises horse owners to consult with homeopathic horse health practitioners for advice on their horse's specific requirements.

To give your horse the correct dose, you will need to advise your practitioner of the following: lactating or not, gender, age, height, body size, breed, feed type, activity, natural behavior, and temperament. Most herbs can be sprinkled on your horse's grain or cubes in dry form but some may also be made into a tea and mixed into feed.

Every horse needs plenty of time to graze, stroll, graze, roll, and graze again, on pure and natural grass.

Simonds believes it's no coincidence that incidents of Cushing's syndrome, insulin resistance, pituitary problems, and "undiagnosed disorders and diseases" are increasing. Many toxins and chemicals, such as polychlorinated biphenyls (PCBs) and phosphorus-containing compounds (PCCs), mimic thyroid or pituitary secretions and either prevent the secretions from entering the body or trigger them to be released when they are not required by obstructing communication between nerves and synapses.

There is no practical way to get rid of all the toxins that our horses (or we) may come into contact with, even if we sequestered ourselves outside of society on untouched land. Simonds suggests that horse owners provide carbon water filters for their horses. "There are places in Pennsylvania and the East where they have been mining for so many years and the toxins and metals are so heavy that there isn't a house around that doesn't have a water filter on it, and there isn't a barn that has one. Those metals combine with fertilizers and toxins in the hay, and now you have chemistry going on that you don't even know about — let alone how to get rid of it."

Three Likely Suspects

Three primary toxins can be found in or on hay:

- Fertilizers (possible nitrogen toxicity) may be bound in the grass structures
- Mold retardants (to make the hay last longer, especially in humid areas of the country) may be found on the hay
- Masked, recycled chemical waste, which may be found in and on the hay and comes from the soil

---

TOXIC HAY

According to Mary Ann Simonds, the number-one way that horses encounter toxins is through the hay we feed them. In Canada and the United States growing hay for animal consumption is an unregulated industry. In fact, legally you can hay a hazardous-waste field and then sell it as feed. "Knowing where your hay comes from and knowing the soils it comes from is important," says Simonds. "If it's a local person across the street and you watch the hay, you probably don't have a problem. If it comes from the feed store, it's very likely that they are not having it tested for any potential toxins."

Become educated about where your hay comes from, and if you have to, choose a browner but less-toxic hay and supplement with high-quality cubes. Simonds believes that we need to begin thinking differently about what a horse can eat, especially when biofuel production is competing for field space with hay crops. "Horses can eat things like peanut plants and moss; they aren't just grass and hay eaters," she says. "We need to move toward a more varied diet for horses."

Testing for Toxins

An increasing number of hay brokers and large hay sellers are now testing hay for toxins, something Simonds attributes to a growing awareness and consumer demand. While they are unable to test for a broad range of toxins, they do check for some of the more common toxins. The one that takes the top spot here is nitrogen.

"Nitrogen toxicity occurs when hay fields are overfertilized," explains Simonds. Fertilizers can be made with toxins that can't be disposed of any other way. "For instance, if you have tailings off a plant or mill, you can sell them or get rid of them by giving them to fertilizer companies. They will process them by adding NPK (nitrogen, phosphorus, and potash) because the only

thing fertilizer companies have to do is to list the NPK on the label; they don't have to list the toxic secondary substances that are included. So right there you have an unknown going into the soil."

Beautiful, lush, green hay can be deceiving: it might look that way because it has been overfertilized. In that case it might be best to choose browner, lower-quality hay than risk causing toxicity in your horse, which can manifest as hives, diarrhea, and anemia, as well as a host of other issues.

There are a few things you can do to detect, prevent, and recover toxic hay. First, make

---

How to Test Hay

If you often need to purchase large quantities of hay, you should core a sample and send it to a lab for testing. Many larger show and boarding barns do this, especially if they receive large trailer loads every month or so.

If your hay needs are much smaller, however, it may not be financially possible to test every load. Here is an easy method of testing for the presence of toxins (although it doesn't identify which ones):

1. Smell the hay. Hay should have a fresh, green smell that attracts horses. Odor is the number-one reason horses reject hay, and this includes hay that smells of "nothing" and is likely full of fertilizers.

2. Put a flake of hay in a tub of water. After a few minutes it should look like a tea, dark green to brown, depending on the hay. If it turns black or grey, this indicates that chemicals and toxins are present.

sure you know that your hay is coming from a safe place. Buy only from reputable sources, and choose local sources if you can. Second, test your hay in whichever manner you deem appropriate.

Soaking Hay

Soak hay to draw out toxins. You can include a capful of chlorine in the water while soaking three to five flakes. Chlorine has gotten a bad rap due mainly to unsafe handling and accidental household-cleaner poisonings, but chloride ions do bind many metals. (Chlorine in large quantities can dehydrate cells, but in small quantities it's safe.) Chloride is the ion that goes back and forth through the cell barrier, pulling out and binding toxins. Many cities add chlorine to their water systems to destroy harmful bacteria — and you can also use treated water for your soak.

Soaking your hay is a good idea because it removes any dirt or irritants that may have gotten on the hay during transport. Hay doesn't need to soak very long; you can finish other chores and feed the soaked hay last.

Offer Free-Choice Blocks and Licks

Without realizing it, many well-intentioned horse owners can actually make their horses toxic from feeding too many extra vitamins and minerals in their feed. Giving a horse the ability to freely choose his minerals through blocks and licks is the best way to avoid overdosing with them. In addition, it gives the horse an unmasked availability of nutrients: that is, not masked by sugar and flavor. If you observe that the horse is eating more of a particular mineral than he should need, you should assess why he feels he might need it.

Cubes and Pellets

Many horse-health experts decry the use of cubes and pellets because they are not as friendly to a horse's digestive tract as hay. In addition, they are offered in a form that is relatively foreign to most horses, whose teeth have evolved to eat flat blades of hay or grass. If you think you might have toxic hay, however, they can be your best option.

If you use hay cubes and pellets, look for the words "toxin free" or "weed free" on the package label.

> ### Toxic Horse?
>
> A horse with toxicity issues may seem lethargic and anemic. You can test for anemia by checking the capillary refill time (CRT), the amount of time it takes for blood to return to blanched tissue of the gum, which indicates how well your horse's blood is circulating. To check your horse's CRT, lift his upper lip, press your thumb against his gum, and hold for two seconds. After lifting your thumb, you should see a white area on the gum; this spot should return to pink in one to two seconds.
>
> Toxic horses may eat wood or bitter plants and weeds. They are often trying to replace something in their diet or remedy a taste or sensation caused by toxicity.

OLD-FASHIONED HAY

A great place to find toxin-free hay is at Amish or Mennonite farms that use traditional methods of farming without toxic chemicals. Their hay tends to be beautiful and clean.

Traditional farmers have a long history of using horse power fueled by grass.

Cube manufacturers buy large quantities of hay and then dry, chop, and compress it into cubes. During the process they are able to add vitamins, minerals, and nutrients. "Many of the better feed companies are running full tests on their cubes to ensure that there are no toxins," says Simonds. They can then make statements such as "weed free" or "toxin free" right on the product packaging, as well as list the nutrient content. "Even though it's not the best choice for the horse's chewing and digestion," Simonds adds, "from a toxin standpoint it may very well be a cleaner feed to use."

READ THE LABEL

Simonds warns horse owners away from buying the cheaper cubes that do not specifically state either what they have tested for or the test results. All animal feed (except hay and forage) must follow strict labeling regulations. Livestock (including horse) and poultry feeds, however, have slightly relaxed guidelines that allow for some ingredients to be referred to by collective names such as "animal protein products, forage products, grain products, plant protein products, processed grain byproducts and roughage products."

DEWORMERS AND DRUGS

We treat our horses with a variety of drugs and dewormers, often because we believe it is necessary. And as consumers we are trained to treat the symptoms rather than investigate underlying causes. Certainly, many different medications are beneficial, and most do not negatively affect the environment if they are used responsibly and disposed of properly.

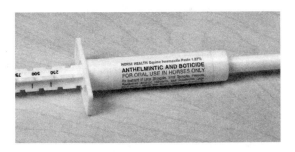

Although many commercial dewormers include no instructions for proper disposal, they are hazardous materials and must be handled with care.

Proper Disposal

Did you know that dewormers are classified as pesticides or miticides? As such, they are considered hazardous waste. The National Ag Safety Database outlines how to dispose of dewormers and drugs properly:

Safe disposal of livestock medicines not only protects the environment, but also you, farm employees, family members, and other livestock from accidental exposure to hazardous chemicals. You may be able to return product or containers to your veterinarian or the manufacturer. In some municipalities, you can dispose of human and animal medicines at "Household Hazardous Waste Days." Animal pesticides (e.g., dewormers, louse control products, etc.) must be disposed of in an identical manner to crop pesticides. Even though recycling is a great idea, you should never reuse livestock medicine containers and should set them aside in a safe place where they cannot be found by children, livestock, and pets until you can dispose of them properly Most clean containers can be disposed of in the appropriate section of your local landfill site (most have specific areas for waste that may include chemicals). Check with your municipality for specific guidelines.

Many horse owners simply toss empty dewormer syringes into the garbage without realizing that they are disposing of a hazardous material. If even a quarter of an ounce is left in a syringe, multiply that by the millions of horses who receive dewormer multiple times per year: that is a lot of hazardous waste entering our landfills.

So what is the proper way to dispose of these syringes?

First, keep them in a sealed container that is used only for this purpose. Call your county Extension or provincial agriculture department to locate the nearest hazardous-waste depot. You can also contact chemical dealers or growers' associations that have pesticide-disposal programs. In some counties in both Canada and the United States, local fire departments accept such waste and other unwanted chemicals for proper disposal several times per year.

Ivermectin Specifics

Ivermectin is a dewormer that is of particular concern to many horse owners. While it breaks down in manure and soil within a matter of hours, it is still considered a hazardous waste, with a toxicity of category IV. Some dog breeds — most notably collie breeds — are very susceptible to ivermectin-related central nervous system toxicity. This tendency traces back to a gene mutation in these breeds that allows the drug to pass through the blood-brain barrier.

"At least twice a summer I see clients whose dogs have died from ingesting ivermectin from manure," says Simonds, who recommends strongly that manure from horses given ivermectin dewormer be bagged and disposed of rather than put in the manure pile. "Horses are not good absorbers and digesters, so much of what goes in comes out. If they have a lot of mucus in their intestines and you give them ivermectin, it won't even get absorbed, so it comes right out again in their manure in a very

THE WORM CYCLE

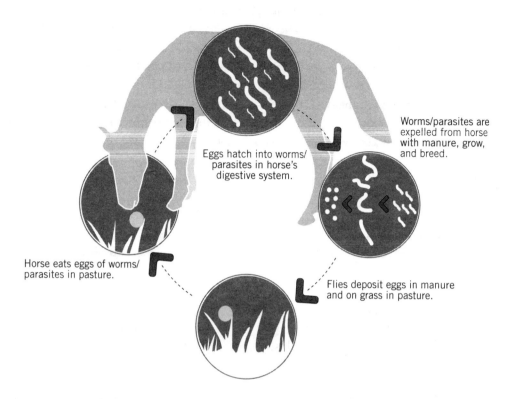

Eggs hatch into worms/
parasites in horse's
digestive system.

Worms/parasites are
expelled from horse
with manure, grow,
and breed.

Horse eats eggs of worms/
parasites in pasture.

Flies deposit eggs in manure
and on grass in pasture.

high concentration, which can get eaten by the dogs." (See Resources for more information.)

Ivermectin has a half-life of 22 to 28 hours and 3 days for its metabolites. It is metabolized by the liver and excreted in the feces over an estimated period of 12 days. Less than 1 percent of the dose is excreted through urine.

Safer Alternatives

So what's the alternative? You can continue to deworm your horse, switching products periodically and sticking to ones that have a shorter half-life. The very reason for this, however, is that worms and parasites are becoming more resistant to the chemicals we use to kill them. A better option for a green horsekeeper is to remove the chemicals from the equation (as much as possible), so that you keep them not only out of your horse's body but out of the earth as well.

Wild Medicines

Medicinal plants that might have functioned as natural dewormers for roaming horses include the following:

- Blackberries
- Periwinkle
- Pine needles
- Queen Anne's lace
- Tarragon
- Wild ginger
- Wild onion

Nature does a great job of controlling worm and parasite infestations on her own through some preventive measures. We've influenced nature's way because our horses don't exist in the wild, where nature has the ability to self-manage. The worm cycle (where eggs are dropped in the manure and then accidentally ingested by another horse) was not a problem when horses were free roaming and wild — they didn't stick around the same pastures. Birds would also follow their movement patterns and eat the larvae; everyone was happy (well, except the larvae). Also there were medicinal plants that helped the horse rid his body of the worms or parasites (see box on previous page).

DE as Natural Dewormer

One popular natural dewormer is diatomaceous earth (DE). This product helps stop all manner of insects through a mechanical pest-control technique that's the equivalent of squashing each bug that comes into contact with your horse. DE is composed of the crushed fossils of freshwater organisms and marine life. Although it feels like a fine powder, when viewed up close, DE resembles cut glass. The particles are far too small to cause any damage to the inside of your horse, as long as it's the medical-grade variety (which is the kind you'll buy when you purchase this at your feed or farm store) rather than the coarser industrial grade. The powder particles, however, are just the right size to be deadly to parasites and worms. They scratch the surface of an insect's body or shell and cause it to dehydrate by absorbing the lipids on its exoskeleton — a little nasty for the bug but wonderful for the horse owner who wants an environmentally friendly alternative to pest control.

DE is also a great alternative for fly control, as it can be spread around a stall or paddock. Be careful, though, because it can also be dangerous for the good bugs you want to keep around, such as predatory beetles and ladybugs.

TRACKING THE FECAL COUNT

Try using these natural methods and monitoring your horse's fecal count to ensure you are making a difference. You may still want to use a chemical dewormer occasionally (maybe once a year) but only if required. In fact, fecal counts should be done whether or not you are choosing a natural deworming method, because you want to ensure that the method you are using is effective.

You can have your horse's manure analyzed by collecting two or three fresh manure balls from the horse to be tested and sending the manure sample to a veterinary laboratory. Results are expressed as eggs per gram (epg) of manure; a count of less than 200 epg suggests a light parasite load that is not a matter for concern. Horses with high fecal egg counts of 500 to 1,000 epg may require further intervention.

NONTOXIC PEST REPELLENTS

Flies are horrible little creatures when they begin to gather on the eyes and underbelly of a horse; some horses are so sensitive that the presence of these winged tormentors can drive them crazy. Other flying insects can carry dangerous diseases.

Dozens of companies market fly and pest repellent products to horse owners. There is, of course, competition for your dollars, with claims about which product lasts longest, kills the most, and contains the newest chemicals. Some companies are marketing fly sheets with chemically infused material.

Horse Pal Fly Trap

There are natural alternatives, such as the Horse Pal Fly Trap. It may not surprise you to know that many insects, such as the biting horse fly, the deer fly, and the yellow fly, are predators. As "blood-feeding flies" (of the Tabanidae family), they are not attracted to baits that rely on chemicals or smells to attract them; instead they hunt with their eyes. The Horse Pal has been designed to attract them, visually, from a long distance. The flies are attracted to the black ball under the trap, believing it to be blood-filled and a potential food source. They enter the capture bottle and are killed from the heat of the sun on the bottle.

This product has been field tested by the University of Wisconsin–Extension Department of Entomology, which reports that "it caught an impressive number of horse and deer flies." Michigan State University also tested the product and found that "during a four-day run in July (a total of 36 hours of collecting) it caught 340 Tabanids (at least 4 different species) for an average catch of 9.5 flies per hour. The best catch rate during that time was 27 flies per hour, the worst was 2.5 flies per hour (on a cool, cloudy day). During the seven-day collecting period in early August when the population was waning, the trap caught an average of 3.6 flies per hour.

The range of the catch rate was from 1.9 flies per hour to 6.2 flies per hour."

The trap will not attract flies that hunt by smell, such as blackflies, houseflies, botflies, and stable flies.

Zero Bug Zone

A product called Zero Bug Zone (ZBZ), recommended to repel mosquitoes, houseflies, termites, hornets, and wasps, consists of an opaque container with a liquid inside that you hang in your barn. Several containers are needed, as it works in a 60-foot (18 m) diameter.

The science and technology behind Zero Bug Zone are difficult to explain and apparently also a well-guarded secret. A nonpoisonous, nontoxic liquid inside the product is charged, and the static charge given off works on the nervous

---

 TAIWANESE MOSQUITO TRAP

It turns out that mosquitoes are a universal annoyance! Inside the barn you can create a no-fly zone with smaller homemade products, such as the Taiwanese Mosquito Trap. Designed by a class in Taiwan for a school project, this mosquito trap has been built and used around the world, thanks to the power of the Internet. (Use your favorite search engine to find photos online.)

MATERIALS NEEDED

- 1 half-gallon (2 liter) bottle
  Knife or scissors
- 1 cup (200 mL) hot water
- 3½ tablespoons (50 grams) brown sugar
  Thermometer
- ¼ teaspoon (1 gram) yeast
  Black paper

WHAT TO DO

1. Cut the top of the bottle. Put the top part aside.
2. Put 1 cup (200 mL) hot water into the bottle, and stir in the 3½ tablespoons

(50 grams) of brown sugar. Put the bottle with the sugar water into cold water to cool it down to 104°F (40°C).

3. After the sugar-water mixture cools down, add the yeast. When yeast ferments, it creates carbon dioxide.
4. Put the top upside down to fit into the bottle, and tape in place.
5. Put black paper around the bottle because mosquitoes prefer their carbon dioxide in dark places.
6. Put the trap in a dark, humid place. The sugar-water-and-yeast solution should be replaced every two weeks.

system of the insect, causing disorientation that encourages it to leave the area.

"The thing people like about this product," says Allan Wells of The Healing Barn in Millbury, Ohio, "is that it's all natural. They don't have to worry about releasing any chemicals into the air that can be absorbed by you or your animals. Many products lure insects and catch or kill them. This product strictly repels them."

It can take five to seven days to create the zone, but Wells reports that some happy customers notice a difference in 12 hours. "We can ride in our arena at night with all the lights on and not one mosquito," says Wells.

If you have spent any time in Mexico, you may have seen restaurants hanging bags of water to repel houseflies. The fly sees the big clear bag as a predator and steers clear. After a few days, however, the water becomes cloudy and must

Hanging bags of water are a common sight in Mexico where they keep houseflies away for a limited period of time. The flies are confused or threatened by the sight of large, round orbs of water and steer clear.

be changed. ZBZ does not become cloudy and will repel houseflies in this manner.

It is important to note that bugs may still enter the "zone," even though they quickly become disoriented and leave. Therefore, the product does not provide absolute protection from West Nile virus. (See Resources for more information.)

Barrier Methods

Let's not overlook the most common barrier methods for pest management: horse sheets and masks. Sheets and masks are both viable options for horses that spend a lot of time outside — mainly because we aren't likely to head out to the pasture multiple times a day to apply fly repellent.

Sheet manufacturers have been coming up with increasingly ingenious ways to get us to buy their products. One of the newest methods is including insect repellent right in the fibers of the sheet. While this is a nice thought, I doubt that the chemicals are required. I suspect they may only serve to increase the chemical challenges your horse will face in a day.

For a more natural route, consider just *not* brushing your horse every once in a while. A horse that is allowed to get a little dirty and muddy has started to create a natural barrier from insects. I know that when you see your horse caked in mud from nose to tail your first thought is the amount of elbow grease that you are going to need to invest in the cleanup. It's more likely, though, that your horse isn't bothered by the mud at all — he's probably relieved to be "bugged" less by the insects.

CLEAN OUT YOUR CABINETS . . .

. . . And your garage, and your tool shed, and your tack room. Horse people can accumulate a lot of chemicals quite easily. We clean saddles and horses and stalls; we maintain trucks and tractors and keep the wheelbarrow wheels turning with a little WD-40 every now and then. But there are many chemicals we can — and should — do without.

If you go into almost any boarding stable or commercial barn, you'll find a bottle of some kind of cleaner under a sink somewhere. That's because as humans we have this insatiable desire to get rid of germs. And as much as we love our horses, we think they have more germs than we do.

Beware of Bleach

So we clean the coffee room with bleach, throw some down a stinky drain, clean out a stall when a sick horse has been in it, and clean the barn bathroom. All-purpose bleach, right? Well, it certainly has plenty of effects.

- There are more poisonings by bleach than any other household cleaner
- Over time bleach breaks down clothing fibers
- When combined with an ammonia product (which perhaps went down that drain earlier in the day), it produces chlorine gas

(a chemical weapon used in World Wars I and II)
- Factories that produce bleach also release carcinogenic dioxin by-products (which have been found in the milk of cows) and mercury (a dangerous element that affects the central nervous system in an exposure)
- When it comes into contact with organic compounds (like, oh, manure), it can form harmful by-products such as trihalomethanes, which are by-products of chlorination and include several chemicals such as chloroform, bromodichloromethane, dibromochloromethane, and bromoform (all are suspected cancer-causing agents)

Try these alternatives to using bleach:
- Chlorine-free products like Seventh Generation Chlorine Free Bleach, Bi-O-Kleen Oxygen Bleach Plus, OxiClean Free Versatile Stain Remover
- Good old hot water and soap
- Sunlight can remove stains from horse blankets; simply wash in cold water, and hang in sunlight to dry with the stain facing the sun

Looking at Labels

If you want to know what's in your chemical-based cleaners, try reading the label. Didn't get very far, did you? Because of legislation in the United States, cleaning companies don't have to put their exact ingredients on the label to avoid revealing a key ingredient — similar to how fast-food fried chicken ingredients are "chicken" and "the Colonel's special recipe." That's why you'll often see "plant based" or "coconut derived" on cleaning-product labels but nothing more specific.

We do know, however, that many chemicals found in cleaners are known as "hormone disruptors." The most well known are nonylphenol

---

Chlorine vs. Bleach

We've used the term "chlorine bleach" so long that the words chlorine and bleach have become synonymous, but they aren't. Chlorine bleach is actually sodium hypochlorite, which is formed from chlorine and sodium hydroxide.

Chlorine is a chemical element that can be used as a disinfectant in small quantities. It's often used to keep pools and drinking water clean and can be used to remove toxins from hay, as we've discussed. It's a chemical and needs to be kept in a sealed container and out of reach of small hands and paws.

ethoxylates, found in most detergents. They cause hormonal problems over the long term and can cause issues with how our bodies produce and use estrogen and testosterone. Chemicals used to clean homes, offices, and barns should be kept separate from any feed or water sources, especially in barns with viewing or coffee areas.

ALTERNATIVE CLEANERS

Just because you're getting rid of chemical cleaners doesn't mean you need to stop cleaning. We still get dirty out at the barn and still need to keep clean. So let's make our own cleaners with items we can find in our local grocery store.

All-Purpose Cleaning Solution

Add one-half cup (118 mL) of borax to one gallon (3.8 L) of hot water to produce a very simple ecofriendly cleaner. Make sure the borax dissolves completely, then spray or wipe any surface that can be washed with regular water (not wood).

Cleaning Your Barn

We don't spend a lot of time cleaning barns — well, okay, we do — but not the type of cleaning that involves products under the sink. Instead we focus on the pitchfork-and-shovel kind of cleaning. But when we do clean, this often involves chemicals. Consider the following scenarios and alternatives:

PREVENT CLOGS. Rather than chemicals to unclog your washrack drain, ensure that everyone abides by a strict policy of picking up manure immediately so it isn't washed down the drain. Or if you do have a large clog, have a plumber employ a "snake" to remove the clog manually.

MICROFIBER. Use microfiber cloths to clean windows and flat surfaces. Microfiber cloths — like the Norwex brand — are reported to remove 99 percent of bacteria. They work with just water and are incredibly absorbent. You can also use them to clean your saddle and silver.

DUST REGULARLY. I know, this is probably the most often overlooked cleaning process in a barn simply because there's always dust in barns. But if you keep the dust and the webs away, you will end up with cleaner tack, clothing, rooms, and air.

PRESSURE POWER. Instead of resorting to chemicals to clean washracks or walls, get a pressure washer. It's a healthier alternative.

Cleaning Horses

We often clean our horses, especially in show barns, with harsh soaps and chemicals meant to "whiten whites." Most horse-cleaning products can be done away with and replaced with a mild shampoo made for humans plus some elbow grease. Using harsh chemicals tends to create a cycle of cleaning, dry skin, dull coat hair, and the application of more cleaning products to bring out the shine. More on this topic can be found in chapter 12.

Cleaning Saddles

Prevention is a great tool when it comes to cleaning saddles. Covering your saddle will not only keep it cleaner longer but will protect it from scrapes and marks from that old barn cat that likes to jump up and make itself a bed on the seat.

A damp cloth is the first step in cleaning your saddle. Avoid a wet cloth or sponge; barely damp is best. You don't want to discolor your saddle with a cloth that is too damp. Microfiber cloths also work well for this because they can be damp and yet feel dry at the same time.

Cleaning silver on the saddle is quite easy. Apply a light coat of toothpaste, allow it to dry, and polish it off with a damp cloth. This job is best done when the silver is not attached to the leather.

Cleaning Glass and Stainless Steel

A secret from any fast-food restaurant or diner along the side of the road is that you can clean any glass or stainless-steel fixture with vinegar. Apply vinegar to a cloth or directly on the surface and wipe. Add ¼ cup (59 mL) of vinegar to 2 cups (0.5 L) of water and pour into a spray bottle for great results.

Microfiber cloth is another great option for cleaning glass. Glass-cleaning microfiber cloths are made by several different companies; one of the best is Norwex. Just spray with water and erase everything from fingerprints to grease to dust.

COMMON CONTAMINANTS

Dealing with chemicals is something we cannot escape from on our farms. If we have any machines on the property or horses that require medication, or anything that requires cleaning, we are likely to come across some kind of chemical. And while they can be safe to use, they are dangerous when stored or used improperly.

Oil

Motor oil is a major contaminant on any farm. Sometimes we park on gravel and don't see the oil dripping from our trucks. Or we don't notice that the tractor has a leak until we back up one day and notice that the gravel where the tractor was parked is sure looking black these days.

You might think that it just stops where it drips, but come rain and snow and melting season, that oil flows toward the closest river, lake, stream, or ocean. Just a drop or two in a well can spoil the entire supply.

SAFE CLEANUP AND DISPOSAL

Since we can't run our vehicles without it, oil will always be around the barnyard, so take precautions when handling it. Park vehicles on a cement pad, and clean up any spilled oil with rags. Dispose of it quickly in a safe manner by taking it to an appropriate area in a landfill or to a local garage when you take your truck for an oil change.

Oil-soaked rags are a spontaneous combustion hazard because as the oil oxidizes, heat is released. If the heat is not dissipated, it can build up and ignite the rags. If you cannot dispose of rags safely and immediately, lay them flat on a high, cool surface. This will allow the heat to dissipate.

Consider having a reputable heavy-duty mechanic travel to your farm to change the oil on your machinery. He will be responsible for the safe disposal of your used oil.

If you are mechanically inclined and do your own oil changes, be careful you do not spill the oil; take the used oil for recycling at a local car dealership, garage, or repair shop.

Chemical Storage

- Don't store chemicals anonymously in unlabeled containers, forcing you to guess by looking at (or worse, by smelling!) what is inside. Store them in their original containers or label them.
- Don't allow random chemicals to collect in closed or sealed containers or small rooms; the gas seepage can create deadly consequences. Very often we keep chemicals in a specific room, building (like a garage), or cupboard to keep them safe, but we need to ensure that they are regularly used or disposed of and that there is adequate ventilation.

Fuel

Because it is used frequently at farms and ranches in all kinds of equipment, fuel is stored in many different ways. Gas cans and gas-powered machinery (such as your lawn mower that sits unused for half the year if you live in northern regions) pose fire risks, as well as contamination risks if any fuel is spilled. One small spill far away from a water source may seem inconsequential, but if you have the same storage facility for years and spill just once or twice a year, eventually fuel will make its way into ground or standing water.

What's the greenest approach to fuel storage? You should store fuel in no-spill Underwriters Laboratory–approved containers only, far from heat and ignition sources, including outlets and electric tools or appliances.

SMALL GAS-POWERED TOOLS

Small gas-powered machines such as your lawn mower are incredibly fuel-inefficient. A Swedish study conducted in 2001 concluded, "Air pollution from cutting grass for an hour with a gasoline-powered lawn mower is about the same as that from a 100-mile automobile ride." While new regulations from both the U.S. Environmental Protection Agency and Environment Canada mean that gas-powered lawn mowers made after 2007 must be more efficient, nothing will be as efficient as not using one. Consider using a manual mower — or your horse — to keep the grass trim. (Do not let your horse graze for 48 hours after a rainfall, however, to avoid severe root damage by hooves.)

Antifreeze

If you live in a northern climate, antifreeze is used almost year-round. Although most of us don't have firsthand knowledge, many departed dogs and cats have learned that antifreeze has a sweet taste they can't seem to resist. Ethylene glycol is the deadly ingredient that quickly dispatches those who consume it. To top it off, antifreeze has commonly been found to contain traces of oil and lead and needs to be treated as a hazardous household waste.

There is really no substitute for antifreeze, but you may want to choose a propylene glycol–based version that isn't lethal. Any antifreeze product, however, should be handled and disposed of by your mechanic. There is no safe way to dispose of antifreeze at a farm or ranch. Watch for bright green drips coming from under your vehicles and machinery, a telltale sign of an antifreeze leak.

Paint

Remember that summer when you painted the fences white? They have faded and peeled and are in need of a coat again. If you're headed back into your garage to use the same paint as last year, you've committed a big enviro faux-pas by keeping the old cans around. Old paints release gases over time, even when you think they are sealed. These gases are neurotoxic and have been known to cause cancer.

On a purely economic scale, if you have a lot of paint left over, you could have purchased a smaller amount. If you can't find anything else to paint and you don't want to throw it away, find a friend or relative who needs something painted. Better to see it used than have it go to waste or pollute your air.

To dispose of old paint, call your city or county garbage-disposal service. Often local fire stations will have annual household-waste drop-off days, and they'll accept such items as paint and other chemicals.

When choosing paint, make sure you choose one that has low levels of volatile organic compounds (low VOC). As discussed in chapter 4, these toxic compounds are not good for your lungs or the environment. The good news is that many new low-VOC paints are readily available for household and farm use.

Fertilizers and Pesticides

Most landowners don't dispose of fertilizers or pesticides at all, believing that what they don't use this year can be kept for the next season. Most people realize these products are poisonous and don't know how or where to throw them away. Consequently, they are responsible for a lot of poisonings each year in children under the age of six.

As an environmentally conscious small- or large-acreage owner, you should not need chemical fertilizers if you are composting your manure properly, as outlined in chapter 7. Organic farmers do not use chemicals at all, or they risk invalidating their certified-organic status. They use quality compost, cover crops (such as nitrogen-rich alfalfa), and crop rotation to nourish soil naturally and to allow it to rest and regenerate. You may need small quantities of fertilizers to increase the nitrogen content of your compost pile, which you can buy at garden centers and dispose of properly. To learn how both fertilizers and pesticides can be disposed of, call your city or county sanitation department.

Medical Treatments

Many horse sports have come under increased scrutiny with respect to the use of drugs, supplements, and nutraceuticals (foods or herbs that may have a medicinal effect). In some big-money sports such as Thoroughbred racing, drugs and chemicals given to horses can have disastrous effects, even if not life threatening. For example, the use of some off-label anti-inflammatories has been linked to infertility — not something you want in a stallion destined for the breeding shed after winning a few million dollars.

Veterinary advancements are wonderful, and it is irresponsible to refuse the advice of your veterinarian simply because a therapy involves drugs or medication. If you do desire to keep your horse in a more natural state, however, discuss nonmedicinal treatments such as herbal or physical therapies.

While the topic of holistic health is outside the scope of this book, the philosophy is the same. Holistic health practitioners view the body as a whole unit and treat it as one rather than treating single symptoms or signs of disease. Similarly, environmentally friendly horsekeeping views the environment as a whole entity and recognizes that what affects one section of the environment then affects the environment of the whole.

☼ ❅ CLIMATE VARIATIONS

There is no area of the continent that is free from chemicals. No matter where you live, there are medicines, dewormers, pesticides, insecticides, and other chemicals that you can come into contact with. Make sure you know exactly where the disposal sites are located in your area and keep up to date with legislation that governs chemical disposal. Don't buy a product without considering what you will need to do with it when it has either expired or been used up.

Subarctic

With such a short growing season in the subarctic zone, many farmers may rely on fertilizers and mold retardants to get as much hay from their land as possible and make it last as long as they can. A dry year can be disastrous if the growing season is affected, meaning you'll have to haul hay in from longer distances.

Humid Continental

High humidity in the humid continental region may mean that mold retardants or salt is used on hay to stop mold growth or to dry the hay before baling. Salting is not as big a concern, but overuse of mold retardants can create unhealthy hay.

Humid Oceanic

Because of the steady year-round temperature in the humid oceanic zone, there is no dormant season for worms, parasites, and pests as there is in other regions, so you must remain vigilant about their extermination and life-cycle management. Perhaps if you had a few colder days in your calendar, eggs and larvae in your horse's manure might have a shorter life span. But no such luck. Guess you'll have to make do with all that warmth and sunshine.

If you are concerned about the amount of deworming medication you are giving your horse because of your parasite-friendly climate, consider an alternative, such as diatomaceous earth (DE), mentioned earlier in this chapter.

Highlands

In the highlands climate region, colder temperatures in the winter months can mean a break from your horses' regular worming and pest-management practices. Most insects are dormant during the winter season or their eggs and larvae are killed by the cold temperatures, so you do not have to take as many precautions until the spring.

Semiarid

Decreased rainfall in the semiarid region can mean that any fertilizers used on neighboring land (or on your own) can become more concentrated in the rainwater. Make sure that any runoff from land that has had fertilizer applied is properly diverted (see chapter 8).

Arid

It's likely that fertilizer use is high in the arid climate zone because of great challenges during the long, hot growing season. Drought may affect the availability of good hay from year to year.

IV
Riding Green

WHETHER OR NOT you own land or plan to in the future, there are things you can do right now to make a difference in the environment. In the following pages you will find new processes to follow, new healthy habits to form, and new ways of viewing your responsibilities as a horseback rider.

SHOWING, BREEDING, AND WINNING THE ENVIRONMENTAL WAY

Considerations for the professional horseman

PROFESSIONAL HORSEMEN and -women who train, show, or breed horses have several challenges as they try to minimize their environmental impact. For them to be successful at their jobs, some concessions may be necessary.

If you train reining horses in North America, for example, there is only one National Reining Horse Association Futurity and Championship show, in Oklahoma City, Oklahoma. No matter where you live, if you want to compete at the highest level in this discipline, a lot of driving may be required, and thus your carbon footprint will be larger than you wish. Similarly, breed organizations may restrict shows from being held too close to one another on the same weekend. This rule is intended to avoid competition for attendees, but it means that you may have to drive farther to attend a show.

There are concessions you can make at home to offset the carbon footprint of those miles of driving and ways that you can be as energy efficient as possible and still be successful.

DRIVING DOWN FUEL COSTS

The number-one challenge of showing horses in an environmentally friendly way is fuel usage. If you are showing at a national level, you can usually qualify for one championship show if you have shown and done well in enough regional events. Attending a championship may mean not only traveling across the country but also many weekends of travel simply to qualify.

Competitors striving to conserve fuel may choose not to attend the larger shows, which in turn may mean decreased entries at the smaller shows because fewer people are trying to qualify for championships. This trend might ultimately lead to a less vibrant competitive horse-show industry, which also happens during recessionary cycles, when the financial strain on competitors shrinks show attendance. We want to be environmentally friendly and minimize our carbon footprint, but we also want a viable horse-show industry.

A COMPETITIVE REINER'S TRAVEL LOG

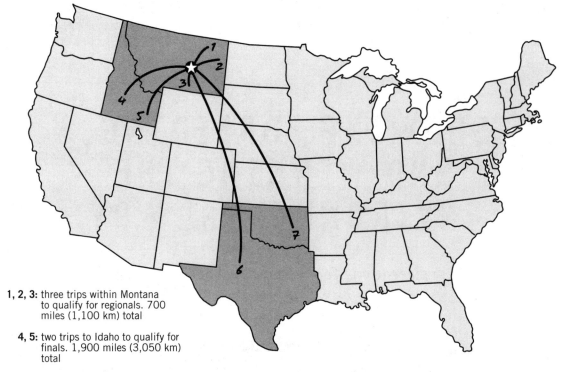

1, 2, 3: three trips within Montana to qualify for regionals. 700 miles (1,100 km) total

4, 5: two trips to Idaho to qualify for finals. 1,900 miles (3,050 km) total

6: one trip to Texas to attend big show. 2,700 miles (4,350 km)

7: one trip to Oklahoma to attend national championship. 2,500 (4,000 km)

This map shows one competitor's driving mileage to and from seven horse shows during one year while qualifying for finals in Oklahoma City. Total: 7,800 miles (12,500 km).

TOP THREE WAYS TO REDUCE FUEL COSTS

1. Carpool. While you can do your best to drive economically and modify your habits, as well as your engine, to be more fuel efficient, the number-one way to conserve fuel is to ride-share. Having more people travel together means less fuel is being consumed, even when you factor in the heavier horse trailers.

2. Stay tuned. Keep your horse trailer in top condition to ensure you are traveling at the most economical rate. Ensuring your trailer tires are inflated to the correct pressure means that they create less pull on your truck and your trailer hauls more smoothly.

3. Lighten up. Hauling lightweight trailers, such as those made of aluminum, has always been my preference. Some feel that those of steel construction will last longer, while others find the aluminum also lasts and is lighter to haul. Either way, it's important to use your trailer in an appropriate manner — there is no sense in hauling one or two horses in a six-horse angle haul.

When oil prices rise, fuel costs explode and become the highest unavoidable expense of showing horses. Just a few horse-show seasons ago, although Canadians and Americans alike might complain about the cost of fuel, they continued showing. In these days of unpredictable gas prices, the reality is that fuel costs may determine which shows you attend in a year.

Diesel is the preferred choice for active, "green" horse haulers. In the past diesel has been priced less at the pump than gasoline, a trend that has recently changed. You will go farther per gallon with a diesel engine, however, because it is more fuel efficient and the new regulations for low-sulfur content make it a healthier alternative.

In October 2006 the Environmental Protection Agency mandated new Ultra-Low Sulfur Diesel (ULSD) regulations that stated the fuel could have no more than 15 parts per million sulfur. In fact, Europe has been light-years ahead of this trend, allowing only 10 parts per million. And now that American diesel comes in at about 7 or 8 ppm, exports to Europe have increased. Supply and demand have influenced the higher price.

Change Your Driving Habits

Hauling horses or simply driving in your truck can become routine, so you aren't considering how you are doing it or what you could be doing differently. Very few drivers take the time to look at their own habits. For example, have you noticed that the drivers going faster than you are all crazy and the drivers going slower are hazards and shouldn't be driving? But what about you?

The tachometer tells you how fast your engine is "revving," or how many revolutions per minute (rpm) it is doing. The higher the rpms, the more fuel is consumed. If you've done any hauling, you will know that downshifting to maintain speed on an incline or to help brake on

---

Change in MPG Ratings

In 2008 the miles per gallon (MPG) ratings that appear on the window stickers of new cars changed. The testing protocols developed in the 1970s were no longer viable for the new millennium because our driving patterns have changed.

Revised estimates will reflect the effects of:

- Faster speeds and acceleration
- Air-conditioner use
- Colder outside temperatures

---

a decline will push your rpms higher. Although higher rpms use more fuel, downshifting may be necessary to drive safely. If you drive an automatic with an overdrive option, the truck may make the decision for you when to shift down or back up again. If you notice that the truck is doing this too often, consider turning the overdrive off and adjusting your speed as required, keeping your rpms around the 3,000 range and your speed at a safe and economical 55 miles (90 km) per hour, as this will stabilize fuel consumption.

To save time and energy, plan your trip ahead of time. Knowing what the traffic conditions are like and taking into consideration possible road construction and rush-hour peaks can save fuel by avoiding a lot of idling time. Many drivers hauling horses prefer to drive around cities rather than through them, though that might increase the distance. But driving a truck at 55 miles per hour for an extra hour is more efficient than stopping, idling, and starting again at traffic lights and crawling through traffic jams for half an hour. Accelerating from a stop, idling at a left-turn signal, and braking suddenly will increase gas consumption. Plan ahead to increase mileage and save fuel.

Diesel

Diesel is a very popular fuel for those hauling horses around the country. Many pickup trucks used for pulling horse trailers are diesel because they are more fuel efficient than gasoline engines. As well, diesel is a by-product of the gasoline-refining process. Along with better mileage, diesel engines produce less carbon dioxide than gasoline engines do. They do produce more particulate matter, however, which is why you will see more exhaust coming from the tailpipe of a diesel truck.

A MORE ECONOMICAL ENGINE

With fuel prices increasing dramatically over the past few years, engine modifications to make a diesel engine more fuel efficient are becoming increasingly popular. And the list of various modifications is long: high-flow air filters, performance spark plugs, electronic tools and gauges, capacitor spark assist, premium ignition systems, vaporizers, lubricants, free-flow exhausts.

The most basic modification is the installation of a computer chip that "remaps" the electronic control unit (ECU) in your engine. (You don't want a cheap one that simply tricks the ECU but one that you can adjust as you require.) The ECU is the brain of your engine, feeding it fuel and regulating airflow. This installation may not increase fuel efficiency in regular driving situations, but it does provide higher efficiency during working situations, such as when pulling a trailer and climbing hills.

The chip modifies the engine by increasing horsepower (the Thoroughbred in your engine) and torque (the workhorse). This enables the engine to run more efficiently when it's towing or when it meets a grade.

To increase fuel economy in diesel engines, add a diesel fuel conditioner year-round. This increases both performance and efficiency by cleaning out the soot and carbon from the

---

Diesel Fuel Mileage and Speed

The *GM Product Service Training Manual for the 6.2L Diesel Engine* outlines diesel fuel mileage for its trucks as follows:

The diesel, like any engine, is affected by driving habits. Speed is more critical on a diesel than a gas engine. On the highway, in the 50–75 mph range, the fuel economy will go down about 3 mpg for each 10 mph increase in speed. A gasoline engine will lose about 1-½ mpg for each 10 mph increase in speed. This condition is perhaps the most significant factor in obtaining good fuel economy. Fuel economy may vary as much as 5 mpg in a given vehicle with different drivers.

---

engine and moving parts. During the winter months this additive is even more crucial, as winter fuel is often less potent than the summer variety.

If you are traveling on a longer trip, consider installing a gauge to monitor exhaust temperatures. The higher the temperature, the harder your engine is working. This helps you by indicating when you should change your driving habits or rest the engine. The harder an engine works, the more likely that it will use oil and other fluids.

Biodiesel

Much ado has been made about biodiesel as a source of "green fuel" in recent years, and there are new vehicles being brought into production every year that can run on biodiesel. In 2007 alone, 450 million gallons of the fuel were produced.

Biodiesel contains no petroleum but is often blended with regular petroleum diesel in order to create a "biodiesel blend." Compression-ignition (diesel) engines can use it with few or no modifications. In addition, biodiesel is simple to use, biodegradable, nontoxic, and essentially free of sulfur and aromatics. It is sold in a mix designated in ratios from B5 to B100. The number represents the percentage of biodiesel in the fuel; for example, B20 contains 20 percent biodiesel.

Creeping up on the radar, however, are claims that the increasing demand for biodiesel is driving up food prices around the world as farmers abandon their food crops and grow crops to make fuel. The National Biodiesel Board claims that it isn't so and that more than 80 percent of soybeans grown are still used for animal feed or human food and that only 5 percent are used to produce biodiesel.

A downside to biodiesel is that it produces more nitrous oxide than do the traditional fossil fuels used to run vehicles. The emissions can be controlled, however, using catalytic converters.

Biodiesel is the new kid on the block when it comes to fuel efficiency. This product is made from renewable resources such as vegetable oil, animal fats, and used cooking oil, and alcohol is used to separate the nonfuel components. The by-products are glycerol (used to make cosmetics, soap, and toothpaste) and seed meal (used in livestock feed). To top it all off, the exhaust of biodiesel smells a bit like popcorn or doughnuts!

HOME-BUILT BIODIESEL REACTORS

Home-built biodiesel reactors: sounds scary, doesn't it? There's an element of risk to any kind of "home-built reactor." Certainly, making your own fuel should not be undertaken without forethought. But it's worth considering. Kit Cosper and his family's home-built biodiesel reactor produces the biodiesel fuel that runs their farm equipment. They gather waste vegetable oil produced by local restaurants, add a solution of methanol and lye (either potassium

STRAIGHT VEGETABLE OIL (SVO)

If you're interested in a humorous take on rising fuel prices, meet Doug Fine, who converted his "ROAT" (Ridiculously Oversized American Truck) from diesel to Straight Vegetable Oil.

The manufacturer Doug used was Plant Oil Powered Diesel Fuels Systems, Inc. (POP Diesel), headquartered in Albuquerque, New Mexico. Founder Kevin Forrest is a vehicle mechanic, U.S. Air Force aerospace maintenance technician, and U.S. Air Force flight engineer with more than 12 years of maintenance and design experience. With his training in aerospace engineering concepts, Kevin conceptualized and designed the POP Diesel Fuel System and clean, safe POP Diesel Fuel. Kevin opened the first state-permitted vegetable-oil fuel station in North America, in Albuquerque, New Mexico, and he's on his way to establishing a network of fueling stations and accrediting installation technicians nationwide.

According to information on the company's Web site, waste vegetable oil is filtered and treated to remove particles, impurities, and water, and special additives are mixed in to eliminate the chance of engine deterioration. The company says that POP Diesel Fuel is clean and safe for your engine. They will ship POP Diesel Fuel directly to you or will work with you to build a filling or fuel-processing station in your area.

Find out more about Doug, his ROAT, and POP fuel systems online (see Resources).

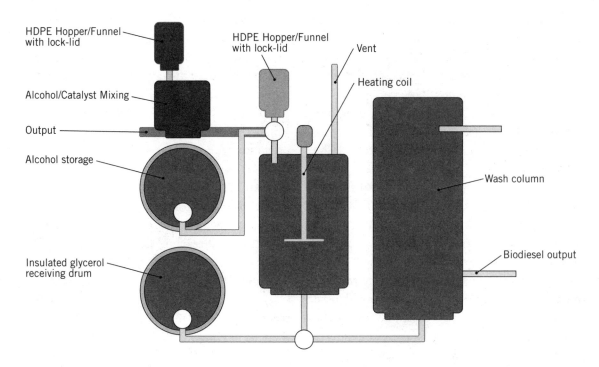

Many inventors are coming up with ways to produce nonpetroleum-based diesel fuel. Here is a scheme for a "biodiesel reactor" like the one built by Kit Cosper. It heats used vegetable oil and a methane-lye solution and ultimately produces biofuel for diesel engines.

hydroxide or sodium hydroxide), and mix to a temperature of about 140°F (60°C).

At that point, what is called "dirty" biodiesel and glycerin results. Next the glycerin is allowed to settle and is drawn off; the resulting biodiesel goes through a series of freshwater washes followed by a drying stage. Excess glycerin from the biodiesel production is fed into a manure pile, where the nitrates break it down completely into benign components. When the temperature drops, traditional diesel is blended in with the biofuel to run the engines.

You can learn more at the Web site of the National Biodiesel Board (see Resources).

Flex Fuel

Flex-fuel vehicles are designed to run on straight gasoline, straight ethanol, or a combination of the two. The ethanol-based fuel is commonly called "E85"; several light-duty trucks on the market can run on this fuel, but not all gas stations carry it. In the United States you can find alternative fueling stations online to plan your trip; unfortunately for Canadians, as of 2008 there are only three (yes, count them, three) stations in Canada that sell E85. (See Resources for more information.)

The E85 option is not the most efficient one for your truck. According to *Road & Travel* magazine, "while E85 burns cleaner, vehicles using it lose an estimated 20 to 25 percent in fuel economy. According to EPA estimates, a two-wheel-drive F-150 FFV gets about 15/19 MPG on gasoline, but 11/14 with E85." Consider using gasoline when pulling a horse trailer and E85 for regular drives to give the environment a bit of a holiday from your gasoline consumption without too much of a hit to your fuel economy.

Fuel from Food

There are growing concerns about the environmental impact of using corn and other food products to make fuel ethanol. Corn in particular requires a lot of fertilizers and pesticides to grow a good crop. Transportation to get the corn to the processing facilities leaves another carbon footprint.

Unfortunately, the math on ethanol fuel does not add up. Depending on the type of cropping system used, 1.5 to 2 pounds (0.7–0.9 kg) of fertilizer nutrients may be necessary to produce 1 bushel (35 L) of corn. It takes 1 bushel of corn to make 2.5 gallons (9.5 L) of ethanol.

The demand for corn and other food products for fuel has had a detrimental effect on food prices as well. Flour prices could rise if farmers grow less wheat and more corn, and corn prices could rise if the demand continues to increase for ethanol.

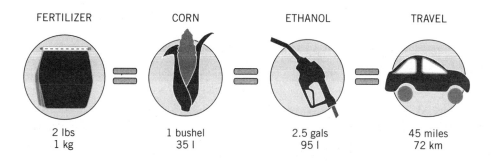

FERTILIZER	CORN	ETHANOL	TRAVEL
2 lbs	1 bushel	2.5 gals	45 miles
1 kg	35 l	95 l	72 km

ULTRA-LOW SULFUR DIESEL AS A VEHICLE FUEL

Ultra-low sulfur content in diesel fuel is beneficial because it enables use of advanced emission control technologies on light-duty and heavy-duty diesel vehicles. The combination of ULSD with advanced emission control technologies is sometimes called Clean Diesel.

Nitrogen oxides (NOx) and particulate matter (PM) are the two most harmful diesel pollutant emissions. These emissions can be controlled with the use of catalytic converters (for NOx) and particulate traps (for PM). However, sulfur — in amounts that used to be allowable in diesel fuel — deactivates these devices and nullifies their emissions control benefits. Using ULSD enables these devices to work properly.

In general, ULSD should cause no noticeable impact on vehicle performance, although fuel economy might be slightly reduced because the process that produces ULSD can also reduce the fuel's energy content. Removing sulfur from diesel reduces lubricity (that is, the capacity for reducing friction). This issue can be resolved by the addition of additives that increase lubricity of the fuel prior to retail sale. In addition, blending biodiesel with ULSD also increases lubricity.

Using ULSD in older diesel vehicles might affect fuel-system components or loosen deposits in fuel tanks. These vehicles should be monitored closely for fuel-system problems and premature fuel-filter plugging during the transition to ULSD. New vehicles designed to use ULSD must never be fueled with a higher-sulfur fuel. If kerosene is blended with ULSD for improved cold-weather performance, it must be ultra-low sulfur (15 ppm or lower) kerosene. New engine oils have been developed for use with new diesel vehicles fueled with ULSD.

Source: the U.S. Department of Energy

LIVING IN YOUR TRAILER

The choice of where to stay when you are on the road can be influenced by your trailer. Purchasing a living-quarters trailer can be an economical choice if you have enough horses to warrant the larger trailer. If you are hauling one horse and a large living-quarters trailer, it may not be efficient, but if you can haul other horses with you, then it becomes more appropriate.

Staying in a hotel is a nice luxury, but depending on the location of the show, you may be driving back and forth several times a day. It's much better to have a green living-quarters space to relax in. Here are two considerations to take into account when looking at environmentally friendly living spaces on the road:

The right size. Do you need a living-quarters trailer or a camper on the back of your truck? On one hand, a living-quarters trailer is more aerodynamic, creating less drag than a camper. However, if you are only hauling one or two horses and therefore only need a two-horse trailer, carrying a camper will add less weight and will be more efficient.

Energy options. How will you power your living quarters or camper? A common choice is to use a vehicle battery to power the smaller lights inside your camper and a larger generator to power bigger appliances such as your fridge or water pump. The good news is that renewable energy is making progress in this section of the industry, with solar power becoming a viable option.

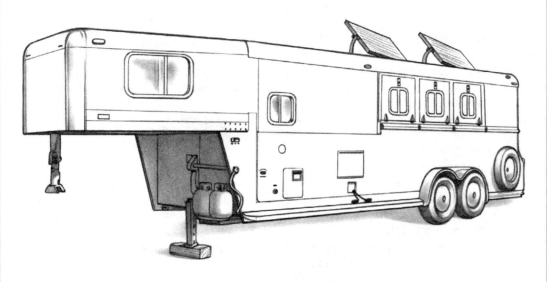

This living-quarters trailer features solar panels to generate electricity.

INCREASING TOW ABILITY

Tow ability is the main concern of horse owners who travel with their horses. It is directly related to fuel efficiency because the better you are able to pull your trailer, the more economical your engine becomes. The harder the engine has to work to pull, the less efficient and economical.

Here are four overall tips to keep in mind:

1. **Slow down.** Regardless of the type of fuel your vehicle takes, how you drive will influence the economics more than anything else.
2. **Reduce wind resistance.** Gooseneck trailers (that attach to a hitch in the bed of your truck) with a tapered nose will help cut wind resistance more than goosenecks with flat or round noses. You can also outfit your truck and trailer with items such as air deflectors and louvered tailgates that are supposed to create better gas mileage, but before you spend the money, make sure that the product is scientifically proven to have some benefit, and make sure you combine it with slower driving.
3. **Balance the load.** With a gooseneck trailer, how you balance the load is not very significant in terms of tow ability, as they are designed to displace 25 percent of their weight on the truck. With a bumper-pull or tag trailer, however, weight distribution is the key to better tow ability and fuel economy.
4. **Add an equalizer hitch.** To maintain efficiency and control, most of the weight in a bumper-pull trailer should be up front. Installing an equalizer hitch allows the transference of weight from the trailer to the vehicle. The hitch is made up of two longitudinal pressure-loaded bars that attach the trailer to the vehicle and help with better weight distribution, which will lead to better fuel economy.

CHOOSING THE RIGHT VEHICLE

Make sure you take a good hard look at your truck and trailer needs. Do you require a bigger trailer but are making do with something smaller? Do you have something larger when a smaller trailer will do? The number of horses and the frequency of trips will factor into the type of vehicle and trailer you need.

A smaller vehicle would in theory be more fuel efficient. But if you are underpowered, using a half-ton truck to pull a trailer that needs at least a three-quarter-ton vehicle, your truck is going to use a lot more gas, and its engine life will be reduced because it has to work too hard.

Truck Choices

Taking care of your rig also means using it in an appropriate manner. Don't force a smaller truck to pull a bigger trailer, as it will put undue strain on its engine, wearing it out faster and using much more fuel than it should. Make sure you have the right tool for the job.

There are several different ratings (see box, "What's Your Weight Rating?," on page 187) that trucks and trailers have for driving, carrying, and towing. These ratings refer to the weight that your vehicle can carry or tow — if this is exceeded, your truck is officially overloaded.

Problems with overloading include:

ENGINE: works harder, consumes more fuel

TRUCK: becomes harder to steer, has longer stopping time, handles differently

TIRES: experience greater wear

ACCELERATION: vehicle may not perform as expected under greater loads

BRAKES: do not work the way they are designed to work at optimal load weight

SHOCK ABSORPTION: wear and tear on frame increases

INSURANCE COMPANIES: may not cover vehicles operating at overloaded capacity

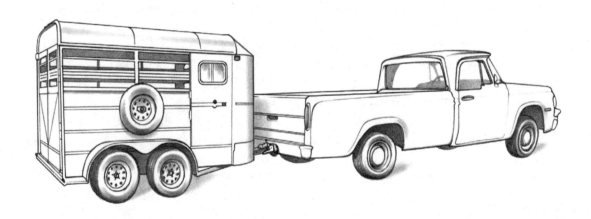

Suit your pulling vehicle to the size of your trailer to maximize fuel efficiency. If you don't require a larger truck, don't pull with one. Conversely, don't try to pull a larger trailer with a smaller truck.

WHAT'S YOUR WEIGHT RATING?

Both trucks and trailers have weight ratings that tell you how much weight you can carry or pull. It is vitally important that you adhere to these weights for both safety and efficiency.

TRUCK

- Gross Vehicle Weight Rating (GVWR) is how much your truck can carry. This includes not just your cargo, but also your passengers and fuel. If you have a camper trailer on the back of your truck, this is an important number to know.
- Tow Rating refers to the maximum weight your truck can haul.
- Base Curb Weight (BCW) is the weight of the truck with a full tank of fuel.
- Cargo Weight is the BCW plus any additional cargo or added weight put into the truck.
- Maximum Loaded Trailer Weight (MLTW) is a very important number: the maximum amount of weight that your truck is rated to haul. You should never buy a trailer that

when fully loaded will exceed this number. Take into consideration how much tack, horses, and living quarters weigh.

- Tonnage Weight is how much weight is pushing down on your trailer hitch. Exceeding this number will put undue stress on your truck's frame.
- Tow Weight is Tonnage Weight plus GVWR and determines how much weight the truck is able to safely pull with a loaded trailer.

TRAILER

- Gross Combined Weight Rating (GCWR) is the maximum combined weight that a truck and trailer can weigh, including cargo, passengers, and fuel.
- Gross Axle Weight Rating (GAWR) is the maximum weight that one axle can carry and includes the weight of the axle. This is an important number to know if you are carrying heavier horses.

Brenderup Real Trailers

When I first saw a Brenderup trailer (pulled by a Volvo station wagon on the Autobahn), I thought the owners of the horses inside were completely crazy. No way would I allow my horse to be inside such a small trailer. However, there is much more to it than meets the eye.

Each trailer is individually constructed with solid phenol resin panels that keep the interior cool — unlike most metal and aluminum trailers, which see an increase in internal temperature as the day gets hotter. The fiberglass roof is one solid piece and is molded aerodynamically for peak fuel efficiency. In addition, these trailers have a specialized braking system that

adjusts for load and speed. There are many incredible features, but is this product "green"?

Very. Brenderup trailers are lightweight and designed to be efficient and create less stress on your vehicle. This translates to less fuel and the ability to use a smaller, more efficient vehicle for hauling. According to the Brenderup Web site (see Resources):

The chassis design limits the amount of weight the trailer puts on the tow vehicle. This makes the required wheel-base-length (of the tow vehicle) unimportant thereby allowing you the freedom to choose from a very wide range of suitable vehicles. You are no longer limited to a big truck when trailering a couple of big horses.

As well as being safe, lightweight, and fuel-efficient, the Brenderup trailer is cool and comfortable for the passengers and can be towed by an average-sized vehicle.

Aerodynamically, the trailer is also superior. Behind other trailers a vacuum is created at speeds of more than 44.6 miles (71.8 km) per hour, which creates two problems. First, it causes sway in the trailer when it is passed by a larger vehicle, increasing strain and pull on the towing vehicle. Second, the vacuum drags on the trailer, pulling it in the opposite direction from the one you are traveling in, making your engine work harder to pull it. To minimize this effect, some manufacturers choose to put more weight on the front end of the trailer, with heavy tongue weight for stabilization. The tow vehicle then needs to be strong and wide to pull the trailer.

But with a Brenderup trailer the sloping roof reduces the vacuum considerably and therefore reduces the need for the manufactured compensation. All that's required to pull this trailer is an engine with a minimum of 120 horsepower and a wheelbase as short as 93 inches (2.4 m). Such vehicles as Jeeps, Dodge Caravans, and Toyota 4Runners are all acceptable.

TEN TIPS FOR FUEL-EFFICIENT HAULING

Following are the "ten commandments" of fuel efficiency.

1. **Drive sensibly.** Speeding decreases your gas mileage. The Canadian Broadcasting Corporation (CBC) released a report in April 2006 that stated, "The Canadian Automobile Association and Natural Resources Canada say it takes 20 percent more fuel to go the same distance at 120 kilometers (75 miles) an hour than it does at 100 km/h (62 mph)." And the U.S. Department of Energy says that every five miles an hour you drive at speeds of more than 60 mph calculates to spending an extra 20 cents per gallon of gas. While you may say that a gallon of gas will not cost you more than you paid for it, if you use it up more quickly driving from point A to point B, then your fuel efficiency, well, tanks.

2. **Take a little off the top.** Roof racks or empty roof storage containers can decrease

your fuel economy by 5 percent because of wind resistance.

3. **Load wisely.** According to Ron Belzil from Trailer Canada in St. Paul, Alberta, "Remember to adjust or balance your load by placing adequate tongue weight on the front of your trailer to insure proper tow ability." This includes loading your trailer in a logical manner so that the horses are balanced, by weight, in the back.

4. **Avoid idling engines.** It's no surprise that idling engines get zero miles to the gallon. During the summer months, there is no reason to leave your vehicle idling. As winter approaches, you need to warm up your gas vehicle for only 30 seconds. Your engine doesn't need to be "warm" to be driven; the purpose of this 30-second warm-up is to heat the oil for better lubrication.

It is commonly believed that you should leave a diesel engine idling for a longer period of time than a gasoline engine. This is not necessarily true, however, because a diesel engine at normal idling speed uses a very small amount of fuel, and it does not actually keep the engine warm. On a cold day at normal idling speed the engine is actually cooling itself down because it's not making enough heat to run efficiently.

You can install a fast-idle switch that will increase the idle speed and actually keep the engine warm. The driver can accomplish this just as well by keeping the truck at a manual high idle or driving a short distance. Technically, your engine is warm enough to operate safely when the gauge has registered any increase at all in engine temperature.

5. **Stay properly pumped.** Underinflated tires can lower gas mileage by 0.4 percent for every one pound per square inch (psi) drop in pressure of all four tires. Imagine running with properly fitted sneakers versus running with big winter boots on — one is going to require more energy than the other. Belzil states, "We recommend filling up all tires on trucks and trailers to about 80% of the manufacturer's recommended maximum because the maximum is the 'do not exceed' point, not the ideal inflation point. You will find this number right on the tire."

6. **Tire tune-up.** While bald tires are a safety hazard, driving on the wrong tires will cause problems as well. If you still have your winter tires on after the snow and ice are gone because you think they will give you better traction, consider that the deep tire treads increase rolling resistance and decrease your fuel economy.

7. **Seek and save.** Try using a free online service, such as www.GasBuddy.com, to find the lowest fuel price in your area. Gas prices will typically increase before a weekend (Thursdays being the most popular day to raise prices, according to gasprice watch.com) so gas-station dealers can take advantage of the increase in demand from weekend warriors. Try filling up your truck on Wednesdays.

8. **Slip into overdrive.** If your vehicle has an overdrive gear, use it. This decreases your engine speed while maintaining driving speed, saves gas, and reduces engine wear.

9. **Oil tune-up.** Each vehicle manufacturer recommends a particular grade of motor oil for your truck that will improve its performance. Look for an "Energy Conserving" label or oils that have friction-reducing additives.

10. **Healthy trucks = happy drivers.** Tune up your truck regularly. Dirty air filters and worn out spark plugs rob the engine of fuel efficiency. Also, faulty oxygen sensors can decrease gas mileage by as much as 40 percent.

PLAYING ECO-HOST

If you own a larger facility and host events on your property, you can be a powerful advocate for environmental friendliness as well as set a great example.

Begin by having recycling bins on your property. Old oil drums or bins that are labeled PAPER, GLASS, and PLASTIC will do fine. Make sure that you mention several times in your entry package, show schedule, and over the loudspeaker that the bins are there to be used.

Other environmentally friendly options include the following:

- Give away refillable water bottles as prizes. Don't sell bottled water — instead, give water away and provide the bottles, such as the stainless-steel Klean Kanteen.
- Ban Styrofoam cups, and charge just 25¢ for a refillable mug of coffee.
- Print the show bill on recycled paper.
- Have all hookups run on solar-powered energy.
- Seek sponsors that are environmentally friendly.
- Have a tack swap
- Use rainwater to control dust in the arena

Contact your local Extension service, conservation district, or municipal district, and ask them if they'd like to have space at your show to set up a booth; they may be able to demonstrate products and instruct spectators on ecofriendly horsekeeping.

When hosting an event at your facility, you have the perfect opportunity to educate other horse owners about the importance of being environmentally friendly. Don't just tell them the barn rules, but explain the reasons behind them. For those who are most passionate about being "green" horsekeepers, you can host seminars and workshops at your barn.

Alayne Blickle from Horses For Clean Water has been hosting workshops in the King Conservation District (Washington) for more than a decade. He teaches mud management, manure composting, and water care to interested horse owners every month and throughout the seasons. You can find more online (see Resources).

INNOVATIVE MANURE MANAGEMENT AT HORSE SHOWS

In Wellington, Florida, showing horses is big business. Between the National Horse Show, the Winter Equestrian Festival, and the Palm Beach Polo Club, more than 100 tons (91 tonnes) of horse manure is produced per day. Unfortunately, the Village of Wellington, in Palm Beach County, does not allow stockpiling of manure, so it must be hauled off-site for composting.

But the surrounding areas can only use so much manure. The Florida Department of Environmental Protection became concerned that runoff from mismanaged manure would contaminate the Florida Everglades, a large shallow-water system with a very diverse ecosystem.

In October 2007 the Wellington Village Council authorized negotiations with Wellington Energy LLC to discuss turning all that manure into electrical energy and selling it to Florida Power and Light. According to the proposal, the power plant would have the capability of burning 5.4 tons (4.9 tonnes) of manure an hour, resulting in 1.5 megawatts of power being fed back into the grid. If the project is approved, it will take about 18 months for it to become fully operational.

BRINGING UP AN ECO-BABY

Raising horses is truly a labor of love. There are not just financial investments but emotional ones as well. To raise the healthiest horses, you need to take some environmental factors into consideration.

Diet

Do you have enough pasture space to support mares and foals? Mares have specific dietary requirements during their 11-month gestation period. You will need to supplement grass with good-quality, toxin-free hay and ensure that water is available at all times. You will have to be extra vigilant about maintaining your pastures to make sure that the grass they eat is of the highest quality.

Young, growing horses and pregnant mares should receive a balanced ration containing sufficient calcium and phosphorus in a ratio of from 1.5:1 to a maximum of 2:1. Ratios of calcium to phosphorus greater than this can cause growth and development problems, such as epiphysitis (an inflammation of growth plates).

You will also have to ensure that your pastures are free from weeds and dangerous plants. Some varieties of tall fescue can contain an endophyte fungus that can cause severe health problems. Mares are especially sensitive to the fungus. During the last three months of gestation, mares should be removed from pastures containing endophyte-infected tall fescue.

Fencing

Once the foals are born, you will need to consider fencing options. Generally, foals stay with their mothers; however, some precocious youngsters do wander and can become separated. They may simply lie down next to a fence, roll over, and find themselves on the other side. I've also witnessed foals getting tangled in electric fence, which can be quite traumatic for them. The fencing you choose should be sturdy and safe; if there's a loop of rope or a loose strand of wire, a foal will find it.

During weaning time each spring, you will need to separate mares and foals. This often requires a field that provides separation between mares and foals. Ideally, the mares and foals will not be able to see or hear each other. If this is not possible due to space issues, then you may need to make separate arrangements for one of the pair to live off the property for several weeks. It is too dangerous to keep the mare and foal separated but still within view of one another, as some foals may try to climb fences to get back to their mothers.

If you intend to keep a stallion for breeding purposes, he also needs special consideration. His turnout area cannot share a fence with another horse because stallions can be aggressive, as you undoubtedly know. Some stallions may require reinforced fencing, in case they decide to take out their aggressive impulses on the nearest board or fence post.

Breeding

Breeding between mares and stallions can be accomplished in several ways. The traditional option is to allow the stallion to live in the pasture with "his" mares. This certainly works. The concern, however, is that mares or stallions may be injured when they are allowed to mingle. This is an even bigger concern than with mares and geldings because a mare may reject a stallion or a stallion may become too aggressive, and it's all fun and games until someone gets kicked in the wrong spot. Also, most breeders choose to keep mares and foals separate from the stallion, as a foal can get in the way at the wrong time and also sustain an injury.

You can also use "live cover" or hand breeding. Either your mare must be trailered to wherever the stallion lives and kept there for several weeks to be bred, or (if you're the stallion owner) you must have the ability to board a mare for the same length of time. This requires additional facilities to house extra horses. Also take into consideration the fuel costs of ferrying horses back and forth between facilities.

Another option is to use "shipped semen." This means that semen is collected from a stallion and shipped in a specially cooled container to wherever the mare lives, even if across the country. While definitely a good option for the convenience of horse owners who want to breed to a horse on the other side of the country, there is the environmental cost of shipping material across the country.

Finally, consider whether you need to breed your horse. If you are a professional and are making a long-term investment in a possible show horse, then you are likely taking many factors and costs into consideration. But if you want to breed horses "just because," ask yourself: With all the unwanted horses going through auctions and even being shipped for slaughter in other countries — do you need to create another horse to feed, or can you feed one that is already here?

☼ ❄ CLIMATE VARIATIONS

Subarctic

If you are considering setting up a training area in the subarctic region, you will be one of only a handful doing it. Because of the low population density in this area, you will have to travel great distances for competitions. You don't have to compete to be a trainer, of course, so consider using innovative communication techniques; for example, your clients can watch their horse being ridden and keep up to date on his progress through webcams or videos.

Humid Continental

The biggest challenge you will face in the humid continental climate zone is trying to arrange a nice, warm, sunny day so you can hold your horse show in the first place. You can try relying on a farmer's almanac to predict the weather, but have backup plans in place in the event of rain or high winds.

Humid Oceanic

There is a long history of horse competition and training in the moderate humid oceanic climate, with sports such as polo and show jumping dominating. This area sees quite a few horse shows each weekend, and facilities will need to compete for entries. Try being unique with a focus on environmental awareness.

Highlands

The beautiful scenery in the highlands zone makes it a great place to live and work. Although you may have to travel a fair bit to attend competitions, more are being established each year in your region. The main challenge is attracting competitors from other regions to come and show despite high fuel costs.

Semiarid

The heart of horse country is in the semiarid region: for instance, Texas has the highest horse population in North America. This is a great area to live in for showing and training horses. You are located centrally to access feed, clientele, and competitions. In fact, your greatest challenge may be an overabundance of trainers around you.

Arid

The biggest problem you will face in the arid climate zone is training during the summer. While it's the hottest time of the year, it's also the busiest show season. You may have to get up earlier to train, take a break from about 11:00 A.M. to 3:00 P.M., and then ride later in the evenings.

SMALL ECO-HOOFPRINTS

Environmentally friendly trail-riding strategies

WHILE YOU ARE BEING SO CONSCIOUS of nature and the environment and doing your best to respect the earth, make sure you take time to go out and enjoy it. One of the best ways is to take your horse trail riding. There are many parks and paths dedicated specifically to trail riding with your horse. Others you'll share with fellow outdoor enthusiasts, such as hikers and mountain bikers.

In recent years efforts have become more common to ban equestrian use of many trails because of damage caused by horses and riders.

HOW HORSES DAMAGE TRAILS

On the trail most riders know to let their horse choose its own path. Horses choose the safest route, often in the footsteps of other horses, and this can make the trail less viable for other users. For example, many hoofprints may create gullies in the middle of a trail. When the trail is muddy, hooves act as suction cups dislodging soil and expanding muddy, wet areas.

While some overnight campers will carry feed, in most trail-riding scenarios horses eat grass and legumes found along the trail and drink from streams and rivers. If horses walk along streams, their hooves may damage the riparian areas (as discussed in chapter 8) by knocking soil into the water and accelerating erosion. Sensitive plants and root systems may be damaged. And horses aren't concerned about where they deposit manure, even in flowing water, causing additional issues for other trail users.

Horses, because of their weight, larger hoofprint, and quantity of hoof-falls, are more likely than other trail users to churn up mineral soil and make a greater impact on vegetation, stream banks, streambeds, and riparian areas. In addition, horses that come from outside the local region have been found to transfer alien weeds and plant species from other regions via their manure.

Some dispute the theory that invasive plants are hitchhiking in your horse's manure. They believe that it would be only by incredible luck that a single invasive seed would germinate in the middle of a trail, where it's much more likely to be destroyed by other trail users before it could take root. Proponents of the theory, however, state that there are other areas where manure falls, especially when users are riding on private land or off regular trails, and seeds are accidentally planted in lush, fertile fields.

This situation may seem harmless at first glance; however, invasive plant species by their nature compete with native vegetation for resources (minerals, nutrients, water). According to the U.S. Department of Forest Ecology and Management:

> Non-native species also have a pronounced economic impact. Weeds cost the U.S. economy $32 billion a year by decreasing crop production by 12 percent, and 73 percent of the weeds are non-native. The cost estimates . . . excluded costs of (i) producing the herbicides ($4 billion), (ii) programs to control non-native plant species ($3 billion), and (iii) environmental and public health damage caused by herbicides ($9 billion). Pastures for livestock are especially susceptible to invasion by native and non-native weeds, with an estimated 45% of the invading species being non-native plants. Forage production from pastures is a $10 billion industry in the U.S. and yield losses caused by non-native species total $1 billion annually.

And those are just the financial costs; the environmental repercussions are even higher, including the consequences of using herbicides. Some of the herbicides in use are known to be mutagenic (causing changes, or mutations, in your genetic code or DNA), carcinogenic (cancer-causing), or teratogenic (disfiguring, especially as related to birth defects). The best approach is to acquaint yourself with the plants that grow (and should not grow) upon the trails where you ride so you can help eradicate invasive weeds before herbicides are required.

Who Has the Greatest Impact?

A trail-impact study from the Aldo Leopold Wilderness Research Institute comparing hiking impacts to horses and llamas noted:

> Horse traffic resulted in statistically significant higher sediment yields (the primary indicator of trail deterioration) than either hiker or llama traffic. The low-level horse treatment (250 passes) caused more impact than the high-level llama treatments (1000 passes), suggesting that horses can cause at least four times as much impact to trails under the conditions simulated in this experiment. In addition, under dry trail conditions horse traffic caused significant reductions in soil bulk density (a measure of how compacted the soil is) compared to llama and hiker traffic. Horse traffic also caused significant increases in soil roughness compared with the other two users. This suggests that the greater impacts of horses on trails is a result of soil loosening of trail surfaces that are otherwise compacted, thereby increasing the detachability of soil particles and increasing sediment yield and erosion.

— *Llamas, Horses, and Hikers: Do They Cause Different Amounts of Impact?* Thomas Deluca (University of Montana) and David Cole (USFS–Wilderness Research Institute), 1998 study (Source: AmericanTrails.org)

What You Can Do

There is a Golden Rule of trail riding — treat the trail as you would have others treat it. With all the different types of travelers on North American trails, each has a pretty good idea of who is causing the most damage. Of course it's never them. Accept responsibility for caring for the trail as you travel upon it and invest your time in annual maintenance of areas that you frequent. The following pages have specific steps that you can take to minimize your impact on the trails you ride.

EIGHT WAYS TO MINIMIZE YOUR IMPACT

Ideally those following behind you should not be able to tell that you are ahead of them. There should be no visible signs of your presence and certainly no damage. Here are some suggestions on how to avoid damaging your favorite trails:

1. **Stick to horse trails.** Ride only on equestrian-designated trails that are properly designed, constructed, and maintained with horses in mind. For example, a trail that is designed for horses will have more stones in it to help stop horses' hooves from sinking deep into the mud or soil and damaging root systems or developing gullies.

2. **Watch your speed.** Ride at a pace that is appropriate for the conditions. Going too fast on wet surfaces increases damage to the trail, especially when your horse has to scramble for footing or cannot choose the safe, dry path.

3. **Choose switchbacks.** To minimize erosion do not ride on sloped, linear trails. These trails, often called "fall lines," will eventually become gullies from rainwater and hoof traffic. A properly designed horse trail will follow the lay of the land, switchbacking across the mountain rather than straight up it as the crow flies. Just as farmers combat erosion by plowing along the contour, horse trails will do the same.

4. **Berm, baby, berm.** If there are no natural contours, they can be built into properly constructed horse trails. The Experimental Forest at Clemson University has experienced success with this technique, using geotextiles and geowebs filled with gravel. Geotextiles and geowebs are permeable fabrics that can be used to separate, filter, reinforce, protect, or drain water from soil. They are made from polypropylene or polyester and come in woven, needle-punched (looks like felt), or heat-bonded varieties. These fabrics are used to build berms that create better footing, prevent erosion, and remain in place as the horses travel across.

Creek beds can be popular spots to stop and have a drink, but take particular care to protect the riparian areas and keep manure out of the water.

5. **Take the high road.** Low-lying areas and bog crossings can deteriorate quickly, especially during wet seasons because the heavy footfalls of the horse's hooves inflict great damage to soil and plant roots. A geotextile-and-gravel setup may solve such problems by redistributing the weight of each footstep and not displacing as much saturated soil.

6. **Keep your hooves dry.** Some riders try to avoid damaging trails by riding along or in the streambed, thinking that this option saves the trail. Instead, they are actually causing further damage to the environment. Walking in and out of streams increases erosion and disturbs the streambed, stirring up silt and dirt and muddying the water. A properly constructed water crossing may include more gravel or rocks (possibly used in conjunction with the geotextile material) to provide better footing for horses while entering, exiting, and crossing streams.

7. **Avoid outlaw trails.** Even in parks and protected areas with constructed trails, many riders believe that it's okay to create their own paths. And so many have done this that "user-created trails" or "outlaw trails" have become a problem. Not only are they not maintained for safety and cleanliness, but they damage the roots, trees, and plants located in protected areas. Often such sensitive areas are not cordoned off because it is assumed that all users know that they should stay on the trails.

8. **Be mindful.** Regardless of whether you are riding on a trail that is sanctioned and protected by a park or trail association, on a friend's land, or on trails at your boarding stable, the onus is on you to evaluate the impact you are making on the area with each step. Keep an eye on the trail ahead of you for soft spots, look back for evidence of the impact you're leaving, and be aware of the trail conditions and how they may be changing.

PLANNING WHERE TO RIDE

There are many different regions in North America, and one state or province may have dozens of different regions, each with unique needs and requirements. For example, within any 1,000-mile (1,600 km) radius in Alberta you might find:

- Mountain trails that have problems with trail erosion due to sloping and lack of earthen reinforcement
- Foothills, which comprise rolling hills (some with very sandy soil), winding rivers, and drastic weather changes in the spring
- Trails along rivers and creek beds that may damage riparian areas
- Flat prairie land with minimal elevations, tree cover, or water formations
- Dry, desert "creek bed"–type riding with sandy footing and rocky land formations

As you can imagine, each trail in those areas (whether official or not) requires immediate "on the fly" evaluation. You may not be able to pick up a book on trail riding and learn everything about every area. But you can learn how to evaluate and assess a trail as you ride so that you can participate in a safe, responsible manner no matter where you are.

How Do You Want to Ride?

The type of trail riding you want to do will determine where you ride. You may choose to go trail riding for a single day or a short overnight outing, or you might want to spend a few days in an area. In the United States there are structured, well-designed trails managed by the Bureau of Land Management, the United States Forestry Service, and the National Park Service. In Canada many trails located on public land (also called "Crown Land") are managed by Parks Canada; others are managed at the provincial level by smaller government groups ("councils"), whose mandate it is to promote the creation, preservation, and management of the trails.

If you have not been trail riding before, it might be best to brush up on trail etiquette by reading a book dedicated to the activity (see appendix B). Some aspects of trail etiquette might be appropriate in one area of the country but not in another. Trails and appropriate equestrian use are influenced by the natural soils, flora, fauna, and type of terrain in each area.

Where Do You Want to Ride?

There are hundreds of regions where you can trail-ride in North America, each with its own mix of history, terrain, climate, and soil and water. Before you ride you must assess the region for its unique trail-riding requirements and challenges.

For example, the granite-based soils in the Sierra Nevada region have supported horse and pack-mule travel in and out of gold mines and along supply routes for a hundred years. Old-timers in the area say that some of those trails look the same as they did 50 years ago. They were built with suitable grades so that water runoff would be managed appropriately to allow the year-round transportation of materials in and out of mines. Maintenance largely involves clearing trees blown down across the trails or rockslides, some catastrophic. Therefore, these trails can accommodate use during wet seasons.

In contrast, trails in some of the forestlands in southern Indiana will develop severely rutted trails if ridden too often in wet conditions because the soil is more easily displaced. Damage caused when soil is soft and saturated can result in long-term disfiguration of the land. You can still see the ruts made by thousands of wagon trains crossing the Platte River on the Oregon and Mormon trails — a hundred years later.

NRT Program

The National Recreation Training (NRT) Program is a great resource for Americans.

The National Trail System Act of 1968 (Public Law 90-543) authorized creation of a national trail system comprised of National Recreation Trails, National Scenic Trails and National Historic Trails. While National Scenic Trails and National Historic Trails may only be designated by an act of Congress, National Recreation Trails may be designated by the Secretary of Interior or the Secretary of Agriculture to recognize exemplary trails of local and regional significance in response to an application from the trail's managing agency or organization. Through designation, these trails are recognized as part of America's national system of trails.

These trails are often maintained by groups of volunteers who donate physical labor to keep the trails clean and safe, host special events, post schedules of events, organize a club to apply for extra funding, and generally advocate for the trail. The NRT has a searchable online database that allows you to find equestrian trails in fifty states. Keep in mind that many of these trails are designated multiuser trails, so you may encounter pedestrians, bikers, or motorized vehicles, depending on permitted uses.

One of the best ways to learn about a new area is to contact the trail riding association in the area. These volunteers intimately know their trails in the area and how to ride them while inflicting the least amount of damage. Some may even arrange guided rides for newcomers.

LEAVE-NO-TRACE PRINCIPLES

The "Leave No Trace" principles were developed by outdoor enthusiasts of all creeds to teach users how to reduce their impacts when they hike, camp, picnic, snowshoe, run, bike, hunt, paddle, ride horses, fish, ski, or climb. There are seven basic principles — or "outdoor ethics" — that you should follow when trail riding.

PLAN AHEAD AND PREPARE

- Know the regulations and special concerns for the area you'll visit
- Prepare for extreme weather, hazards, and emergencies
- Schedule your trip to avoid times of high use
- Visit in small groups; split larger parties into groups of four to six
- Repackage food to minimize waste
- Use a map and compass to eliminate the use of marking paint, rock cairns, or flagging

TRAVEL AND CAMP ON DURABLE SURFACES

- Durable surfaces include established trails and campsites, rock, gravel, dry grasses, and snow
- Protect riparian areas by camping at least 250 feet (75 m) from lakes and streams
- Good campsites are found, not made; altering a site is not necessary

In popular areas:
- Concentrate use on existing trails and campsites
- Walk single file in the middle of the trail, even when it's wet or muddy
- Keep campsites small; focus activity in areas where vegetation is absent

In pristine areas:
- Disperse use to prevent the creation of campsites and trails (in other words, where there are no trails and campsites, do not create them by going over the same path or setting up permanent campsites — change your path and rest spots if you visit more than once)
- Avoid places where impacts are just beginning

DISPOSE OF WASTE PROPERLY

- Pack it in, pack it out. Inspect your campsite and rest areas for trash or spilled foods; pack out all trash, leftover food, and litter.
- Deposit solid human waste in catholes dug 6 to 8 inches (15 to 20 cm) deep at least 250 feet (75 m) from water, camp, and trails; cover and disguise the cathole when finished
- Pack out toilet paper and hygiene products
- To wash yourself or your dishes, carry water 250 feet (75 m) away from streams or lakes and use small amounts of biodegradable soap; scatter strained dishwater

LEAVE WHAT YOU FIND

- Preserve the past: Examine, but do not touch, cultural or historic structures and artifacts

- Leave rocks, plants, and other natural objects as you find them
- Avoid introducing or transporting nonnative species
- Do not build structures or furniture or dig trenches

MINIMIZE CAMPFIRE IMPACTS

- Campfires can cause lasting impacts to the backcountry; use a lightweight stove for cooking, and enjoy a candle lantern for light
- Where fires are permitted, use established fire rings, fire pans, or mound fires
- Keep fires small; use only sticks on the ground that can be broken by hand
- Burn all wood and coals to ash, put out campfires completely, then scatter cool ashes

RESPECT WILDLIFE

- Observe wildlife from a distance; do not follow or approach them
- Never feed animals; feeding wildlife damages their health, alters natural behaviors, and exposes them to predators and other dangers
- Protect wildlife and your food by storing rations and trash securely
- Control pets at all times, or leave them at home
- Avoid wildlife during sensitive times: when they are mating, nesting, or raising young, or during the winter

BE CONSIDERATE OF OTHER VISITORS

- Respect other visitors, and protect the quality of their experience
- Be courteous; yield to other users on the trail
- Step to the downhill side of the trail when encountering pack stock
- Take breaks, and camp away from trails and other visitors
- Let nature's sounds prevail; avoid loud voices and noises

Source: Leave No Trace Canada, a national nonprofit organization dedicated to promoting and inspiring responsible outdoor recreation through education, research and partnerships (LeaveNoTrace.ca)

THE ECOLOGICALLY RESPONSIBLE RIDER'S RULES

Just as with any endeavor, there are tenets that you can follow, commandments, even. These are the rules laid out and agreed upon in principle by all who follow that group. The following rules apply to all ecologically responsible riders.

1. LIGHTEN YOUR LOAD

Take only as much horsepower as you need on a trail ride, including pack horses if traveling overnight. Keep equipment and tack to a minimum, exercising judicious selection of what is really necessary and what can be left behind.

This rider (above) has packed with economy and efficiency. These riders (below) have brought an excess of gear.

2. LIMIT HORSE POPULATION AND PARTY SIZE

Many federal agencies are concerned with setting regulations in place that control but do not overly restrict the population of horses and people on the trail. If you head out on the trail with your family — Grandma, Grandpa, kids, grandchildren — you are traveling as one family unit, and somewhere around seven horses is not that bad. But if a local horse club sponsors a trail riding event and 60 riders show up, 60 horses traveling the same path is a very different scenario. Responsible trail riders will ensure that larger groups break into smaller groups of about 10 each and stagger their rides.

3. RESPECT EACH UNIQUE TRAIL AND KEEP IT UNIQUE

Each trail has unique traits you should take into consideration, whether you are one rider or 60. Some trails — such as old logging or

To make sure you are all riding with the same philosophy, talk to the members of your group to make sure they all know the rules. This is especially important if you do not know everyone you are riding with.

mining roads — might be able to handle six riders riding abreast, while others with heavier or clay soils or boggy terrain might only be able to handle single-file riding.

The trail corridor needs to be able to accommodate all the riders in your party; if the corridor can't accommodate everyone, there will be impacts on the side slopes: the upslope and the downslope. If it's too wet and muddy, horses will navigate around the boggy spots and create "braided trails" — three or four paths winding around the main path. Other horses then deviate from the main trail, and pretty soon there are many trails where there should have been just one. This frequently occurs in climates with heavy snowfall when riders navigate around a snowdrift, thereby creating a secondary trail that is not maintained or cleared and has no drainage provision for snowmelt. Resolving this issue means educating all people who are using the trails on how important it is to try to stick to the original cleared and maintained trail.

If you do not respect the trail and ride in a manner appropriate to its terrain, you will damage the trail. Each time you're on a trail, you need to ride in the manner appropriate for *that* trail on *that* day, not the trail you rode last week or the same trail you rode yesterday that was dry and is now wet.

4. WATCH THE WEATHER

In wetter climates (including northern, heavier-snow areas), there's a significant impact when you move horses through terrain after a rainfall. Your horse will have a greater impact on the soil and fauna no matter what precautions you take. You need to stay off the soil until it has had a chance to stabilize. Except in areas with granite or rocky soils, wet conditions will result in a greater impact by every activity: human,

Riding in a group can be a lot of fun; make sure, however, that you are not traveling along the trail in one large herd. Adjust the breadth of your group to fit the trail. A large logging road could accommodate five riders across, but a smaller forest trail may be passable only in single file.

wheeled, or equestrian. The responsible equestrian considers climatic effect on trail conditions, including rain, snow, and wind. Either delay the trip or skip the areas that are wet.

5. BE KNOWLEDGEABLE ABOUT TRAIL OBSTACLES

When a responsible user comes to a trail obstacle, he will have to consider what the best strategy is. One choice is to turn around and go back. Another option is to move the obstacle if possible.

Before moving the obstacle, however, you need to be aware of what the trail regulations are (and who sets them). "It's not just a matter of jumping off and rolling the rock down the hill," says Dr. Mylon Filkins, an equestrian expert with American Trails. You may need to consider other issues and possible consequences. It may be better to leave the obstacle and report to the appropriate ranger or land manager that there is an obstruction on the trail, because you may not be aware of other factors that might affect the removal.

For example, if there is a small rock or earth slide across a trail, you might consider scooping it out. After all, it's only a small slide and you brought a shovel for a reason, right? But what happens if you have not scooped the slide out correctly? Now there is the opportunity for an even bigger slide because the upslope was not stabilized.

Trail crews are trained to remove obstacles and barriers in a safe manner that ensures continued trail safety. The responsible rider has contact with land management before using the trail to find out what is appropriate to do about downed trees or slides across the trail.

6. VOLUNTEER AND GIVE BACK

Almost all public trails in North America benefit from volunteer worker bees organized by the users of the area. By far the two groups who most often participate are mountain bikers and equestrians, perhaps because both groups perceive that they have an impact on the trail and seek to give back by cleaning and restoring damaged areas. Hikers, on the other hand, tend

As animals naturally know, single file is the way to go when traveling in the true wilderness.

to feel that they are contributing by having the least impact in the first place.

By volunteering for maintenance and restoration projects, equestrians become stakeholders and participants in the trail system rather than mere users. They are also more readily involved (and have more credibility) when land management decisions are considered, particularly when those decisions concern trail closures and changes in trail classifications that may ban horses from the trails.

7. MINIMIZE OFF-TRAIL TRAVEL

Everyone wants unique and unrestrained trail-riding experiences. Everyone wants to ride up to the top of a ridge and look out over mountains or ride out into a meadow to get a better picture of the trail scenery. But if we all do that, we create problems. If everyone rides up to the top of the ridge, soon a vertical path exists that will erode after the next rainstorm. This does not mean you cannot ever step off the path, but off-trail use has to be minimized and done in a considerate manner when absolutely necessary.

Some areas are restricting equestrian use to designated trails only, which immediately brings into question how big-game hunters will be able to go off trail to pack out deer or elk or how fishermen who ride to remote locations will be affected. What about when an individual horse strays off the trail or breaks a tether and wanders off? Will these users be ticketed or fined?

There is some argument for free and unrestricted use, specifically related to backcountry travel. Some would say that unrestricted use is appropriate for areas where there is a legitimate and historic use but that it needs to be done with some knowledge of the terrain and taking into consideration other users. For example, switchbacks are built into trails not just to provide the safest route up a mountain but to prevent the soil from eroding from use. But some trail users (usually hikers) cut through a switchback,

heading either directly up or directly down the slope. This again creates a possible erosion issue and is an off-trail journey that is neither necessary nor legitimate.

Backcountry trail riding is not only on managed lands but through wilderness areas that may be protected by legislation or other landowners. To ensure continued access to these types of land, seek approval prior to riding (if necessary) and always ride in a responsible manner.

8. BE CONSIDERATE OF ALL FAUNA, INCLUDING FELLOW TRAVELERS

For equestrians the sound of horses moving down the trail is pleasant. Other trail users, however, including hikers and mountain bikers, do not necessarily agree. One of the main nonequestrian-trail-user complaints is that our horses leave annoying manure piles in the middle of the trail. Scatter any manure your horse leaves behind by stepping off and dispersing it with your boots. This is not only considerate toward hikers and other users, but it goes a long way to countering the belief that horses are responsible for spreading invasive weeds through their manure, as it is much more difficult for a seed to germinate when it's not ensconced in a ball of manure.

Who's Sharing the Trail with You?

National statistics establish the following numbers of trail users in the United States: 73.3 million hikers, 43.1 million single-track mountain bikers, and 4.3 million horseback riders

Sources: Outdoor Industry Association
2003 Participation Study and
the American Horse Council

OVERNIGHT CAMPING

I cannot imagine anything more peaceful than exploring the wilderness on horseback and camping under the stars. Many others feel the same way. As our society becomes more aware of the environmental impacts from many different industries, we simultaneously try to take as much advantage of our green spaces as we can and look critically at the effects others have upon them.

Long before we worried about "environmental impacts," horses were making paths through the wilderness on the plains, mountain passes, and wetlands of North America. As urbanization spread, however, movements arose to set aside areas of our land for "the use and enjoyment of the people." Notice that "use" is not a four-letter word, as many think it should be — it's not wrong to use our natural resources. To use something is far different from abusing it; we can use the land for enjoyment as long as we take care not to abuse it.

When camping in the backcountry or in established horse camps, all the important points listed above apply to the trek in and back out again. Staying overnight, however, increases your interaction with the ecosystem you are enjoying, so greater effort to maintain its cleanliness and viability is required.

Here are some ways you can do that.

Restrain Your Horse Properly

Do not tie your horses to trees unless absolutely necessary and then only for a couple of minutes. "This has been a big no-no for a long time," says Dr. Mylon Filkins. "Horses will circle the trees and paw; I've seen where horses have pawed a huge saucer around the base of the tree and exposed the roots to damage. When we come across this we try to fill them back in. But the next year you'll see the same thing around the same tree — especially if it's near a really good fishing hole."

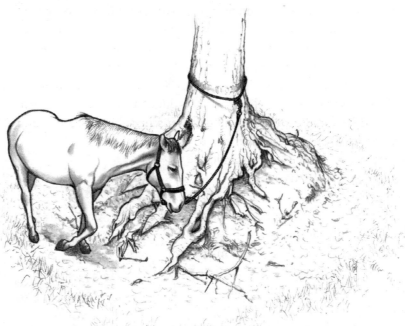

Horses that paw at roots and trunks can cause considerable damage to well-established, beautiful trees. Make sure that your horse is trained to stand without pawing or keep him hobbled.

Turning your horses loose to graze may or may not be allowed. If it is allowed by trail management, there is normally a time limit of one or two nights. Some trails (as in some southern California parks) only allow pasture grazing during certain times of the year, when the soil is stable enough. If grazing is allowed, use your best judgment; do not allow your horse access to boggy, wet pastures.

How Not to Lose Your Horse

Keep pack animals at least 200 feet (60 m) from streams, lakeshores, trails, and camping areas. This helps keep water clean, protects the soil and plants, and keeps trails and campsites clear of loose stock. Rotate stock throughout the area to reduce trampling and prevent overgrazing. Take everything with you when you leave.

SECURING YOUR HORSE

Highline. A highline is one of the easiest, most lightweight strategies for keeping your stock in camp; it also prevents stock from trampling roots and chewing bark. It is easier to put up with a tree-saver strap.

Tree-saver straps. Used with highlines, these adjustable nylon straps will help keep your stock from girdling trees. Here's how to use them.

- Choose a hard, rocky spot
- Place the tree savers and rope about 7 feet (2 m) above the ground
- Stretch the line between two trees using tree-saver straps.
- Run the rope between the straps, tie with a quick-release knot, and pull tight.

Hitching Rails. If you must tie stock to a hitching rail or dead pole, fasten a 4–6 inch (10–15 cm) round pole to two trees at chest height of your horse. Not so high that he will walk under it, not so low that he will step over it. Use rope or twine instead of nails or wire to attach the pole to the tree, and place padding or wooden shims under the lash ropes to protect the bark.

Picket ropes and pins. Take along an easy-to-move picket pin; metal versions are ideal. Avoid areas with obstacles so the rope doesn't get hung up. If you walk your horse or pack animal to the end of the rope before turning him loose, he

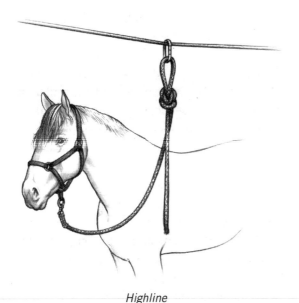

Highline

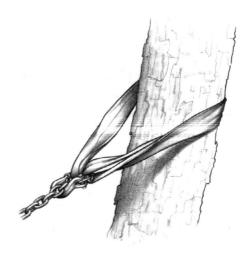

Tree-saver

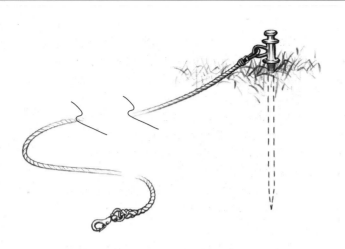

Picket rope

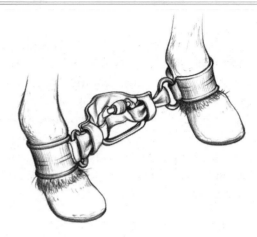

Hobbles

will be less likely to injure himself by running past the end of the rope. Choose ropes with natural fibers to prevent chafing or rope burns, and move the picket pin frequently to prevent trampling and reduce overgrazing. When you break camp, be sure to take that picket pin with you.

Hobbles. Wandering horse? Hobbles work for some animals, but others can move fast while wearing them. Get your stock used to them before going into the backcountry.

CONTAINING YOUR HORSES

If you plan to spend several days in one spot, a temporary corral or fence is a good way to keep your stock in camp. Before leaving home, make sure your animals are trained to stay in temporary corrals. If you find permanent corrals at trailheads or designated horse camps, use them! Try the following temporary fences and corrals (don't forget to take them out with you!).

Plastic snow fences. This fencing and its stakes are lightweight, easy to pack, and come in colors like orange, green, and black. Snow fences can stand on their own and be used with wooden stakes that are easily woven through the webbing or be secured to trees with thick rope. (The reason you can use ropes here is because they are not attached to a moving, fidgeting animal that is confined to one spot to paw at the ground and pull at the rope.)

Rope corrals. Rope corrals are relatively easy to rig and move, but they do require extra rope. One method uses two parallel ropes tied with loops or bowlines and threaded with cross ropes for a more secure enclosure.

Manure and Weeds

As mentioned above, many believe that horses are responsible for introducing invasive or alien weed species into the backcountry. The evidence is disputed, but considering that horses leave behind material that tends to fertilize plants, the perception is there. And we can't deny the possibility.

This is one topic where a little bit of politics comes into play. Some nonequestrian trail users do not want horses sharing their trails, and the argument that they bring invasive species into the area is used as a justification for keeping horses out.

Here's how it may happen. A horse eats pasture, feed, or forage that contains weed seeds. These seeds pass through his system and out of his body with the manure. If the manure is left undisturbed, it is possible that a seed will germinate and grow because it's nestled snugly inside a warm, fertile ball of manure.

Most manure left on the trail is eventually kicked apart and dispersed (which you can do yourself, or horses that follow may do it for you) from the middle of the trail. But you cannot guarantee that manure piles left in the middle of a lush pasture are completely void of weed seeds unless you have taken the precaution of providing weed-seed-free food at least 48 to 72 hours before your ride. Feeding this way is also wise from a veterinarian's point of view, according to Dr. Filkins, to avoid an abrupt diet change on the trail. Some established camps may even offer manure-disposal systems. And since many horses defecate near their horse trailers or soon after entering the trailhead, you can easily remove any manure close to the trailhead before you leave.

Water Wisely

You have two choices for watering your horse when you are on the trail. Take the horse to the water or bring the water to the horse. The most ecologically friendly way is to pack a bucket and bring water to the horse when you stop. If there is any chance that your horse is going to leave manure or urine or track anything into the waterway, don't risk entering the water. Not only will manure floating downstream leave a very bad impression, it's highly unsanitary for others using the water.

If the bank is nearly flat or very solid or rocky, you can take your horse to the water, but ensure that you remove anything he might leave behind and do not allow him to paw at the ground or water or cause damage to sensitive banks. It's best to choose an area far from human crossings or watering areas.

☼ ❄ CLIMATE VARIATIONS

Because trails across North America are incredibly diverse, the techniques you will use to practice ecological trail riding may vary, even along one trail. It may depend on the conditions that particular day, the weather the previous week, or the riders who went ahead of you.

Subarctic

It may be difficult to go trail riding for much of the year in the subarctic climate zone, because of heavy snowfall and ice buildup. If you are hardy enough to ride while snow is still on the ground, be sure your horse has proper shoes with caulks for traction. Follow trail rules and

---

Transitioning to Weed-Seed-Free Food

To eliminate the possibility of transporting invasive weeds into nonnative areas, you need to transition your horse to weed-seed-free food 48 to 72 hours before your trail ride.

1. Locate a certified weed-seed-free seller through HayAndForage.com or the North American Weed Management Association (see Resources)
2. Six days prior to the ride, introduce 25 percent of the new feed with 75 percent of the current feed
3. Five days prior, increase the feed to 50 percent of the total
4. Monitor your horse for any skin irritations (indicating a possible allergy or reaction to the feed) or changes in bowel movements (onset of diarrhea or compaction)
5. Four days prior, increase the new feed to 75 percent of the total
6. Three days prior (72 hours), transition to 100 percent new feed
7. After the ride, you can transition back to your regular feed over approximately two days

regulations for riding during the spring, after the thaw. It may be some time before many fields can be used because the ground will still be saturated from the melting snow.

Humid Continental

In the humid continental zone you will need to check with trail management to ensure you know the rules and regulations for the trail and to determine which areas are off limits. If you are traveling away from home but still in the same region, you may not be aware of what weather conditions were like in the previous 24 hours.

Humid Oceanic

If you are traveling in the humid oceanic zone, it's likely that you will encounter a wide range of both flora and fauna on your trip because this region has such a diverse abundance. Get to know the types of plants you might encounter on the trail so you know if they are protected or they are not supposed to be there. Reporting noxious weeds is a service to all who use the land — any weeds that take root in this region have plenty of sun and water to keep them healthy.

Highlands

When you are riding trails in the highlands climate zone, be aware that higher elevations may take longer to dry out in the spring and summer. Take care on sloped paths, watch for signs of erosion, and report any you see to trail management.

Semiarid

When riding in the semiarid climate zone, know the animals you may encounter on the trail and whether they are dangerous. Some snakes in this region can be deadly.

Arid

Taking some water with you on your trip in the arid region will be very important, even if you are heading out for just the day. Ask riders returning from trips or ask trail management what the current water situation is — it can be dangerous to take a trip into a drought-stricken area.

LIVING "GREEN" is not just a fad or something that is neat or interesting to do. It's a new global movement that is having far-reaching, positive effects. We might not know the difference we are making in terms of cooling down our planet, and we may not be able to see the results in terms of the carbon we're saving or the CO_2 we're not releasing. But make no mistake: Every small step we take toward being "greener" horsekeepers is a step in the right direction. Let's ensure that our children and their children will be able to experience the same joys of horse ownership that we do while living in harmony with our earth.

READER'S GUIDE

I hope you've enjoyed learning about becoming a greener horsekeeper. There are so many actions that we can take each day (big and small) that will help us conserve energy and take care of our environment. The following questions may help stimulate discussion at a group or individual level.

Part I

1. What "green" habits do you already have? How did you come by them? In many cases, older generations (those who lived in Dust Bowl or post–World War II years) have carried on habits of frugality. Have any older generations influenced your behaviors?

2. Which climate zone do you live in? Have you ever lived in any other zones? What are the particular challenges and/or benefits from living in your zone? What temperature range do you think your horse might naturally be most comfortable in?

3. While you may find the thought of eating a horse distasteful, look at it from a different point of view: What are the environmental implications of having a surplus of horses in North America? Currently, only the United States forbids horse slaughter while Canada and Mexico do not. What are the environmental costs of unwanted horses being shipped to other countries?

4. In the past, legislators have considered re-classifying the horse from a "livestock" animal to a "pet" or "companion" animal to give the species better protection from abuse. What are the implications of this sort of change? Consider that as the owners of livestock, we benefit from state/provincial and federal support (and money), possible tax breaks, limited liability laws, and a looser definition of humane treatment. Is that last one ever a good thing? Perhaps it is if you are required to provide your "pet" with overnight stabling in a heated barn when that might not be healthy, and he actually requires only an outdoor paddock and a shelter. Should we treat horses the same way we treat cats, dogs, and hamsters?

5. Horses have left a great fossil record behind for us to track their movement through the ages. What are your thoughts on the

horse's evolution? Theorize why the North American fossil record comes to an abrupt stop 10,000 years ago (extinction, exodus, massive climate change, cataclysmic event).

6. When do you think the horse has most enjoyed (or enjoys) living?

7. Imagine if we all still rode horses rather than owning automobiles — how could we be more environmentally friendly and how could we avoid the ecological disaster that faced cities in the late 1800s?

Part II

1. Considering your climate zone, what renewable energy source do you think would provide you with the most energy?

2. Can you identify "sinks" or "sources" of CO_2 on your property or where you ride? How many of these are naturally occurring and what could be done to protect the "sinks"?

3. Do you use any forms of renewable energy now? If so, what has been your experience? What would you do to improve the method of harnessing this energy? There must be countless ways we can improve on green energy. Solar panels have been around for decades, but in late 2008 12-year-old William Yuan of Beaverton, Oregon, created a 3D solar cell that can harness both visible and ultraviolet light. His cells actually absorb about 500 times more light than a conventional solar cell and 9 times more than advanced 3D solar cells.

4. Do you have any bats in your region? This may affect your choice for wind power. The University of Calgary has recently shown that bats' lungs overinflate as air pressure drops near the fast-spinning wind turbines bursting blood vessels and capillaries. You may want to contact Bat Conservation International to find out about the bats in your area: www.batcon.org.

5. Which renewable energy source do you believe is most "untapped"? Why?

6. What is the history of your land? Do you know of any contamination points? Can you identify anything that was built for convenience on your property where the environmental cost was not taken into consideration? What can you do to fix it?

7. You may not currently own land, but you can draw your ideal property and fill out your "land use wish list" located on page 44. What is your primary land use and how does this affect the rest of your list?

8. Except in some very dry, hot regions, almost every property has issues related to mud. Where are your problem areas, and what is causing the mud?

9. How do you think your choice of fencing affects your horses? Are some fences better psychological barriers than physical ones? What do you consider to be your ideal fencing options?

10. Do you have enough storage space on your property? What items or products would you store more of, if you only had the space? Can you think of some creative storage solutions?

11. Have you ever considered the embodied energy in the structures you currently have? Would that knowledge have affected your choice of materials? Now when you think about where your barn materials will go in the future, what ideas do you have for recycling and reusing them? Are there other items on your property that can be recycled or already have been once? Send me your thoughts to heather@heather-cook.com and you can be entered into a drawing!

12. Do you know if you have any pressure-treated wood on your property? If so, what are your plans for it?

13. Have you ever seen a green roof, either on a barn or in a city? What do you think of green roofs? Would you consider one for your barn, arena, or house?

14. Do your buildings require insulation? Could you do without it and keep the barn warm more naturally?

15. The London 2012 Olympics has an ambitious environmental plan (see page 74). Do you think more large events (from the Olympics to large horse shows) should have a commitment to being green? How can you let them know about your beliefs?

16. Have you ever considered bulk purchases? Or if you already purchase in bulk, have you considered buying more with a friend or family member? What are the challenges and benefits? Does this apply to more than feed and consumables? What about larger-ticket items?

17. Do you know the proper way to clean up a Compact Fluorescent Light Bulb? If not, see page 95, and post a how-to in your barn.

18. Is a portable toilet an environmentally friendly choice for your barn? What are the cost savings in water use? What are the environmental concerns with chemicals or transportation costs?

Part III

1. Detail the current manure management plans for your property. Do you have any plans for increasing the number of horses you have? How will this affect your current manure management plan? Are you currently using the best option for your land? If not, what is involved in changing your processes?

2. What is the market for manure in your area? Do you have any of the following facilities nearby: research facilities, farmers' markets, vegetable and fruit growers, landscapers, gardeners, schools?

3. Do you have a well on your property? If so, what is the age, depth, and health of your well? Are you currently maintaining it correctly?

4. Do you know your water regulations? In Colorado it is illegal to have a rain barrel because you do not "own" the water that falls on your property. What are your thoughts about this?

5. Have you considered multispecies grazing? Do you know anyone who does this? What are the specific challenges of horse-cow, horse-sheep, horse-goat grazing?

6. Could you become involved in a carbon credit program? If you live in the eastern states, contact the Regional Greenhouse Gas Initiative (RGGI), and if you live in the West, try the Western Climate Initiative.

7. Have you created a hay "tea" to test a flake of hay for toxicity? What was the result?

8. What are your thoughts on holistic horse care?

9. Have you tried any natural pest control methods? Which have worked best for you?

10. Regularly inventory your sheds, garages, and storage spaces for any chemicals or products that have passed their expiration date. Make a systematic plan to properly dispose of them. Ahead of time, research the disposal sites in your area so that you will know where to take the items that are out of date or no longer required. Second, resolve to keep all chemicals separate. This is not a "throw it all in the trash bag" type of project. Third, the disposal process is easiest to accomplish in the spring or summer when you can ensure your sheds, garages, and storage spaces are well ventilated. Fourth, wear gloves and a mask.

Part IV

1. Chart the trips you took last year while competing at horse shows. Calculate your vehicle's fuel consumption and determine how many gallons (liters) of fuel you consumed. What are your thoughts on the resulting number? What might you do differently this year?

2. Which driving habits are you able to change? For example, can you let your diesel engine idle for a shorter time? Can you modify anything on your vehicle for better gas mileage and efficiency?

3. There is a controversy regarding the use of corn for fuel. Proponents say it is a viable "green" fuel. Opponents say that harvesting it releases too much CO_2 and that it's contributing to starvation in developing countries as well as an increase in food prices domestically. What are your thoughts?

4. Do you know of a horse show that has made a commitment to be environmentally friendly? If so, let me know (heather@ heather-cook.com) and we can talk about sponsorship opportunities!

5. Do you consider yourself an ecofriendly trail rider? Where do you normally ride? Have your habits changed since reading this book? Who is in charge of maintaining the trails you frequent?

6. Are you aware of any weed-seed-free hay sellers in your area? What are the costs of their feed relative to the current hay you are feeding? What are your thoughts on the benefits of feeding it before a long trail ride?

BIBLIOGRAPHY

Aadland, Dan. *101 Trail Riding Tips: Helpful Hints for Backcountry and Pleasure Riding (101 Tips)*. The Lyons Press, 2005.

Anthony, David W. *The Horse, the Wheel, and Language*. Princeton Press, 2007.

Bach, David. *Go Green, Live Rich*. Broadway Books, 2008.

Card, A. B., and J. G. Davis. " Composting Horse Manure in Dynamic Windrows, no. 1.225." Colorado State University Cooperative Extension, July 2002.

Carson, Rachel. *Silent Spring*. Mariner Books, 2002.

Chamberlin, J. Edward. *Horse: How The Horse Shaped Civilization*. BlueBridge, 2005.

Daly, Judi. *Trail Training for the Horse and Rider*. Alpine Blue Ribbon Books, 2004.

Freed, Eric Corey. *Green Building & Remodeling for Dummies*. Wiley Publishing, 2007.

Hill, Cherry. *Cherry Hill's Horsekeeping Almanac*. Storey Publishing, 2007.

Hill, Cherry. *Horsekeeping on a Small Acreage*. Storey Publishing, 2005.

Hogue-Davies, Vicki. *Horse Trails: The Traveler's Guide to Great Riding Getaways (Coast to Coast)*. BowTie Press, 2006.

Jarymowycz, Roman. *Cavalry from Hoof to Track*. Greenwood Publishing Group, 2007.

McDilda, Diane Gow. *The Everything Green Living Book*. Adams Media, 2007.

National Renewable Energy Laboratory. "Small Wind Electric Systems: A U.S. Consumer's Guide." U.S. Department of Energy, March 2005.

Paige, Jessica. "A Guide to Composting Horse Manure." Washington State University Cooperative Extension, Whatcom County.

Pavia, Audrey. *Trail Riding: A Complete Guide*. Howell Book House, 2005.

Poe, Rhonda Hart. *Trail Riding: Train, Prepare, Pack Up & Hit the Trail*. Storey Publishing, 2005.

Sellnow, Les. *Happy Trails: Your Complete Guide to Fun and Safe Trail Riding*. Eclipse Press, 2004.

Shim-Barry, Alex. *The Environment Equation*. Adams Media, 2008.

Snell, Clarke, and Tim Callahan. *Building Green*. Lark Books, 2006.

Stoyke, Godo. *The Carbon Buster's Home Energy Handbook*. New Society Publishers, 2006.

Suzuki, David, and David Boyd. *David Suzuki's Green Guide*. GreyStone Books, 2008.

Taylor, Nancy H. *Go Green*. Gibbs Smith, 2007.

Trask, Crissy. *It's Easy Being Green*. Gibbs Smith, 2006.

Turner, Chris. *The Geography of Hope*. Random House, 2007.

Venolia, Carol, and Kelly Lerner. *Natural Remodeling for the Not-So-Green House*. Lark Books, 2006.

Wilder, Janine M. *Trail Riding*. Western Horseman, 2006.

Wilson, Alex, and Mark Piepkorn, eds. *Green Building Products*, 3rd ed. New Society Publishers, 2008.

Yarrow, Johanna. *How to Reduce Your Carbon Footprint*. Chronicle Books, 2008.

RESOURCES AND BEST PRACTICES

GOVERNMENT AGENCIES

United States

U.S. Department of Agriculture
www.usda.gov

U.S. Environmental Protection Agency
www.epa.gov

Cooperative State Research, Education, and Extension Service (CSREES)
U.S. Department of Agriculture
www.csrees.usda.gov

The National Climatic Data Center
National Oceanic and Atmospheric Administration
www.ncdc.noaa.gov/oa/ncdc.html

Natural Resources Conservation Service
(formerly the Soil Conservation Service)
U.S. Department of Agriculture
www.nrcs.usda.gov

Canada

Agriculture and Agri-Food Canada
www.agr.gc.ca
Home of the Prairie Farm Rehabilitation Administration (PFRA) and the National Land and Water Information Service

GRANTS AND INCENTIVES

Federal Programs: U.S.

Environmental Quality Incentives Program (EQIP)
National Resources Conservation Service
www.nrcs.usda.gov/PROGRAMS/EQIP
  The Environmental Quality Incentives Program (EQIP) was reauthorized in the Farm Security and Rural Investment Act of 2002 (Farm Bill) to provide a voluntary conservation program for farmers and ranchers that promotes agricultural production and environmental quality as compatible national goals. EQIP offers financial and technical help to assist eligible participants install or implement structural and management practices on eligible agricultural land.
  EQIP offers contracts with a minimum term that ends one year after the implementation of

the last scheduled practices and a maximum term of ten years. These contracts provide incentive payments and cost-shares to implement conservation practices.

Persons who are engaged in livestock or agricultural production on eligible land may participate in the EQIP program. EQIP activities are carried out according to an environmental quality incentives program plan of operations developed in conjunction with the producer that identifies the appropriate conservation practice or practices to address the resource concerns. The practices are subject to NRCS technical standards adapted for local conditions.

EQIP may cost-share up to 75 percent of the costs of certain conservation practices. Incentive payments may be provided for up to three years to encourage producers to carry out management practices they may not otherwise use without the incentive. However, limited resource producers and beginning farmers and ranchers may be eligible for cost-shares up to 90 percent. Farmers and ranchers may elect to use a certified third-party provider for technical assistance. An individual or entity may not receive, directly or indirectly, cost-share or incentive payments that, in the aggregate, exceed $450,000 for all EQIP contracts entered during the term of the Farm Bill.

Walk a Mile in My Boots
United States Department of Agriculture

The "Walk a Mile in My Boots" initiative is a work-exchange program between agricultural producers and government employees. NRCS is working with the National Association of Conservation Districts to implement this partnership program that was developed by the National Cattlemen's Beef Association and U.S. Fish and Wildlife Service. Exchanges will provide opportunities for producers and NRCS employees to learn more about each other's lifestyles, issues, and operations.

This is a national program, so any producer or NRCS employee may apply to participate.

Producers will be signed up as Earth Team Volunteers and will visit NRCS field offices, state offices, or NRCS headquarters in Washington, D.C. They may shadow biologists, managers, or other specialists; conduct outdoor field activities; attend agricultural meetings and work with USDA officials. Outdoor activities might include water control monitoring or wildlife habitat planning.

Government employees will visit a ranch or farm and shadow producers in their daily operations. Activities might include branding and vaccinating calves, moving and feeding livestock, irrigating cropland, building and maintaining conservation buffers, or haying.

The length of the exchanges is flexible, depending on the availability of the rancher and NRCS employee. The average exchange will be five to ten days.

Producers and NRCS employees interested in applying for participation in the program should contact the Earth Team National Headquarters:

USDA-NRCS
888-526-3227
www.nrcs.usda.gov/feature/volunteers

Database of State Incentives for Renewables & Efficiency (DSIRE)
North Carolina Solar Center and the Interstate Renewable Energy Council
www.dsireusa.org

DSIRE is a comprehensive source of information on state, local, utility, and federal incentives that promote renewable energy and energy efficiency.

Sustainable Agriculture Research and Education (SARE)
U.S. Department of Agriculture
www.sare.org

SARE is a program of the U.S. Department of Agriculture that functions through competitive grants conducted cooperatively by farmers, ranchers, researchers, and ag professionals to advance farm and ranch systems that are profitable, environmentally sound, and good for communities.

Online Grant Management System
Western SARE
http://wsare.usu.edu/grants

SARE grants are used to increase knowledge about sustainable agricultural practices and to help farmers and ranchers adopt those practices. The Western SARE program administers grants in several categories that help it achieve those aims. Each grant operates on an annual cycle and is selected through a competitive process.

Applicants are typically informed whether their project has been approved for funding within six months of the submission deadline. Dispersal of funds rests on congressional budget decisions.

Choose your grant

Research and Education (R&E) grants

Also known as Chapter 1 grants, they average around $150,000 in size but can be smaller or greater, depending on the proposed project. R&E grants fund projects that typically involve scientists, producers, ag support agencies, non-profit organizations, and others in an interdisciplinary approach. Requests for pre-applications are issued in April. Following the scrutiny of a technical review, the best pre-applications are asked to submit full applications, due in November. The Western SARE Administrative Council, aided by staff and highly skilled and knowledgeable subject matter experts, selects projects for funding early the following year.

Farmer/Rancher (FRG) grants

These are conducted by agricultural producers, with support and guidance from a technical adviser. Individual farmers may apply for up to $15,000 and a group of three or more farmers may apply for up to $30,000. Producers use their grants to conduct on-site experiments that can be shared with other producers. Projects may

also focus on marketing and organic production. A technical review is held in January, and the grant awards are announced in the spring.

Professional + Producer grants

These grants are similar to those for Farmer/Rancher Grants, except that an agricultural professional, such as an Extension educator or NRCS professional, coordinates the project, and farmers or ranchers serve as technical advisors. A technical review of proposed projects is held in January, and the grants are announced in the spring.

Professional Development Program (PDP) grants

Also known as Chapter 3 grants, these are designed to help university and other agricultural professionals spread knowledge to producers about sustainable concepts and practices. Applicants may seek up to $30,000 for one-year projects and $60,000 for two-year projects in a single state or locale. For regional or multistate projects, applicants may seek up to $60,000 for one year and $100,000 for two years. Applications are reviewed in January, and projects selected for funding are announced in the spring.

Graduate Fellow Grants in Sustainable Agriculture

These grants are funded up to $25,000 to assist graduate students in pursuing their advanced degrees. The grant may last for up to two years and a student may apply for only one grant during the course of his or her study. Applicants must be enrolled full time (according to the institution's requirements) at an accredited college or university in the western region. Applications are submitted in the spring and reviewed in the summer, and grants are awarded in the fall.

Building Better Rural Places
Federal Programs for Sustainable
Agriculture, Forestry, Conservation, and
Community Development
http://attra.ncat.org/guide

> Searching for the right federal grants program? This extensive directory of federal programs for sustainable agriculture, forestry, conservation, and community development was compiled in 2004 by U.S. Department of Agriculture agencies working together for sustainable rural development, in collaboration with the Michael Fields Agricultural Institute and the National Center for Appropriate Technology. Contains summaries of each program, contact information and grant-writing tips.

For more information (or hard copies of Requests for Application) on the available SARE grants contact:
Western Region: *http://wsare.usu.edu*
Southern Region: *www.southernsare.uga.edu*
Northeast Region: *http://nesare.org*
North Central Region: *www.sare.org/ncrsare*

Federal Programs: Canada

Agriculture and Rural Development, Alberta
www1.agric.gov.ab.ca

> Home of the AESA Form Based Component that provides extension grants to support integrated environmental support integrated environmental planning, technology transfer and extension activities, and farm resource management by farmers and ranchers. Agricultural service boards, agricultural and environmental organizations, and First Nations groups are eligible to obtain grants. Applicants are required to prepare three-year Environmentally Sustainable Agriculture (ESA) plans outlining priority issues, program details and activities, partnerships and budgets. Priority management areas are based on local issues and are: Nutrient Management, Integrated Crop Management, Grazing and Riparian Management.

Office of Energy Efficiency
Natural Resource Canada (NRC)
www.oee.nrcan.gc.ca
The NRC offers several grants:
- ecoENERGY Retrofit
- Energy-efficient equipment
- ENERGY STAR qualified products
- Energy-efficient new homes
  If you are building an energy-efficient apartment attached to your barn, you may qualify for grants and rebates.

Nonprofit Foundations

David Suzuki Foundation
www.davidsuzuki.org

> This science-based Canadian environmental organization works to protect the balance of nature and our quality of life, now and for future generations.

MANAGING YOUR PROPERTY

Information & Technical Assistance

Cooperative State Research, Education, and Extension Service (CSREES)
United States Department of Agriculture
www.csrees.usda.gov

ATTRA – National Sustainable Agriculture Information Service
www.attra.ncat.org

> ATTRA is managed by the National Center for Appropriate Technology (NCAT) and is funded under a grant from the U.S. Department of Agriculture's Rural Business-Cooperative Service. It provides information and other technical assistance to farmers, ranchers, Extension agents, educators, and others involved in sustainable agriculture in the United States. (ATTRA was formerly known as the "Appropriate Technology Transfer for Rural Areas" project.)

Land Management

Natural Resources Conservation Service
United States Department of Agriculture
www.nrcs.usda.gov

The National Association of Conservation Districts (NACD)
www.nacdnet.org
> NACD is the nonprofit organization that represents America's 3,000 conservation districts and the 17,000 men and women who serve on their governing boards. Conservation districts are local units of government established under state law to carry out natural resource management programs at the local level. Districts work with millions of cooperating landowners and operators to help them manage and protect land and water resources on all private lands and many public lands in the United States.

Soil Conservation

ATTRA – National Sustainable Agriculture Information Service
http://attra.ncat.org
> List of alternative soil-testing laboratories in the United States

Canadian Soil Information System
Agriculture and Agri-Food Canada
http://sis.agr.gc.ca/cansis
> Home of the National Soil DataBase (NSDB).

Water Management

Canadian Water Resources Association (CWRA)
www.cwra.org

Drink Tap
American Water Works Association
www.drinktap.org

Horses for Clean Water
www.horsesforcleanwater.com
> Horses for Clean Water offers ways to care for horses that improves the farm they live on and reduces non-point pollution. Techniques such as mud management and composting manure offer ways to care for animals that benefit the animals, the farm, the owner, the community, and the environment — all win-win-win solutions.

Incinolet
Research Products/Blankenship, Inc.
www.incinolet.com
> Incinerator toilets

The Low Impact Development Center, Inc.
www.lowimpactdevelopment.org
> This nonprofit organization is dedicated to the advancement of Low Impact Development technology. Low Impact Development is a new, comprehensive land planning and engineering design approach with a goal of maintaining and enhancing the predevelopment hydrologic regime of urban and developing watersheds.

Make Your Own Rain Barrel
City of Bremerton, Department of Public Works and Utilities, Water Conservation
www.cityofbremerton.com/content/ sw_makeyourownrainbarrel.html

National Rural Water Association (NRWA)
www.nrwa.org

North American Lake Management Society (NALMS)
www.nalms.org

Texas Water Development Board
Innovative Water Technologies
www.twdb.state.tx.us/iwt/iwt.htm
> Information on rainwater harvesting, including *The Texas Manual on Rainwater Harvesting.*

CLIMATE AND SKY

Weather Maps & Data

USDA Plant Hardiness Zone Map
The United States National Arboretum
http://usna.usda.gov/Hardzone/ushzmap.html

Renewable Resource Data Center
National Renewable Energy Labatory
www.nrel.gov/rredc
> Provides access to detailed resource information through tools, reports, maps, and data collections, including the Wind Energy Resource Atlas of the United States.

The National Climatic Data Center (NCDC)
National Environmental Satellite, Data, and Information Service (NESDIS)
www.ncdc.noaa.gov/oa.html

Environment Canada
www.ec.gc.ca
> Canadian weather data and climate change information

Solar/Lunar Maps & Data

Dynamic Maps, GIS Data, & Analysis Tools
National Renewable Energy Laboratory
www.nrel.gov/gis
> U.S. solar maps

Astronomical Applications Department
U.S. Naval Observatory
http://aa.usno.navy.mil
> Sun and moon data

CanmetENERGY
National Resources Canada
http://canmetenergie-canmetenergy.rncan-nrcan.gc.ca
> Solar maps of Canada

Understanding and Reducing Greenhouse Gases

Greenhouse Gas Sinks and Sources
Canadian Cattlemen's Association
www.jpcs.on.ca/biodiversity/ghg/booklet
> While focused on cattle and ranchers, this well-written guide breaks down the terminology associated with greenhouse gas emissions and the agriculture industry.

Climate Change – Greenhouse Gas Emissions
United States Environmental Protection Agency (EPA)
www.epa.gov/climatechange/emissions
> Provides greenhouse gas information and a calculator to determine your carbon footprint

GREEN POWER

Alternative Fuels & Advanced Vehicles Data Center
www.afdc.energy.gov
> Alternative fueling station locator

Brenderup
www.brenderup.com
> Trailers with a green design

European Small Hydropower Association
www.esha.be
> Small hydropower plant building guide

"The Kung Pao Smokescreen"
Dispatches from the Funky Butte Ranch
www.dougfine.com/2007/03
> Doug Fine's blog chronicles his journey converting his "ROAT" (Ridiculously Oversized American Truck) from diesel to Straight Vegetable Oil

National Biodiesel Board
www.biodiesel.org

National Ethanol Vehicle Coalition

www.e85fuel.com

    Maps of E85 stations in the United States and Canada

Plant Oil Powered Diesel Fuel Systems, Inc.

www.popdiesel.com

Solar Energy International

www.solarenergy.org

    An association focused on developing solar thermal and photovoltaic technologies

BUILDING

U.S. Green Building Council

LEED Green Building Rating System

www.usgbc.org

    The U.S. Green Building Council is a 501(c)(3) nonprofit community of leaders working to make green buildings available to everyone within a generation. Home of the Leadership in Energy and Environmental Design (LEED) certification program.

    The (LEED) Green Building Rating System encourages and accelerates global adoption of sustainable green building and development practices through the creation and implementation of universally understood and accepted tools and performance criteria.

Greenroofs.com

www.greenroofs.com

Building Products and Materials

Aeroseal

www.aeroseal.com

    An air-duct diagnostic system that finds leaks in ductwork

American Formulating & Manufacturing

http://afmsafecoat.com

    Manufactures Safecoat paints and primers; seals garage floor so spills wipe up easily

Big Ass Fans

www.bigassfans.com

    High-velocity fans for barns and arenas

DirtGlue Enterprises

www.dirtglue.com

    Environmentally friendly, high-tech soil treatment that helps stop erosion

Forest Products Laboratory

U.S. Forest Service

www.fpl.fs.fed.us

    ACQ timber and galvanized nail information

LifeTime Composites LLC

www.ltlumber.com

Low Energy Systems

www.tanklesswaterheaters.com/index.html

    Tankless water heaters that heat quickly and on demand

MSDS online

www.msdsonline.com

    Material safety data sheets (MSDS)

National Pesticide Information Center (NPIC)

http://npic.orost.edu

    Information on CCA and non-CCA treated wood

Solar Shade Systems, Inc.

www.solarshade.ca
> Window shades that can be custom-fit to stop 90 percent of thermal heat gain and UV rays

SolaTubes

www.solatube.com
> Natural lighting for all types of buildings

Stall Skins

http://stallskins.com
> Stall flooring alternative to rubber mats

Wood Treatment Products, Inc.

www.eswoodtreatment.com
> Manufactures EnviroSafe Plus, an alternative for pressure-treated wood

WEED MANAGEMENT

North American Weed Management Association (NAWMA)

www.nawma.org
> Provides weed free forage information and standards

Goats Eat Weeds

Ewe4ic Ecological Services
307-654-7866
www.goatseatweeds.com
> Information on controlling weeds with goats

National Pesticide Information Center (NPIC)

http://npic.orst.edu

Weed Control Equipment & Supplies

Flame Engineering, Inc.

www.flameengineering.com
> Manufacturer of the famous Red Dragon hand-held flamer as well as alfalfa flamers, row-crop flamers (2- to 8-row kits), and a grapevine berm flamer that can also be used in orchards. A major supplier of liquid propane accessories to the flame weeding industry. See the online book, Agricultural Flaming Guide.

Forevergreen Chemical Free Weed Control

www.chemfree-weedcontrol.com
> North American distributor of the Swiss-made Eco-Weeder, an infrared thermal weeder heated by a propane flame passing over a ceramic casing. Models include handheld and push-wheeled weeders for use around the home and in gardens, parks, market gardens, small farms, and orchards.

The Green Hoe Company, Inc.

www.greenhoecompany.com
> Mechanical weed removal

Kimco Manufacturing, Inc.

www.kimcomfg.com
> Mechanical weed removal

M.K. Rittenhouse & Sons, Ltd.

www.rittenhouse.ca
> Heat weeders

OESCO, Inc.

www.oescoinc.com
> Supplier of the Aquacide hot water weed control equipment system, which is geared to nursery production, landscapes, and park departments

Peaceful Valley Farm Supply

www.groworganic.com
> Organic farm equipment and supply dealer, carries: handheld flamers, backpack frames for propane tanks, row crop flame kit suitable for mounting on a toolbar and flaming 4 rows

Sunburst, Inc.

www.thermalweedcontrol.com
> Thermal weed control equipment

Waipuna USA

www.waipuna.com

Waipuna, from New Zealand, specializes in a hot foam system; the foam is derived from coconut sugar and corn sugar and is approved for organic production. A single-burner generator covers a width of 8 to 10 inches. A double-burner generator covers a width of 24 to 32 inches. Currently these are geared to municipalities, park departments, airports, and institutional settings. An agricultural unit is under development, with an aim toward orchards, vineyards, and similar agricultural applications.

Weed Badger Division

Town & Country Research & Development, Inc.

www.weedbadger.com

Mechanical weed removal

GREEN LIVING AND HORSEKEEPING

The Green Guide

National Geographic Magazine

www.thegreenguide.com

Online or hardcopy guide to all things "green" for consumers. Tips and advice can be applied to almost any building or type of land in North America. Very inclusive.

Green Living Ideas

www.greenlivingideas.com

Alternative and renewable energy sources and sustainable living. This Web site provides ideas, tips, and information to help you improve the environmental sustainability of every aspect of your life: home energy, green building and remodeling, cars, food, waste recycling — and everything in between. Green Living Ideas includes information on how to live greener in over 200 different aspects of life.

Mystic Horse

http://mystichorse.com

Web site of Equine Ecologist/Behaviorist Mary Ann Simonds

Safe 2 Use

www.safe2use.com

Ivermectin information

Wind and Hydropower Technologies Program

U.S. Department of Energy

www.windpoweringamerica.gov

Information and consumer guides on small wind power

Composting Horse Manure

Composting in Whatcom County

Washington State University Extension

www.whatcom.wsu.edu/ag/compost

Information and articles on horse manure and on farm composting.

Agricultural Analytical Services Lab

PennState College of Agricultural Sciences

www.aasl.psu.edu

Sample compost analysis report

EnviroHorse

California State Horsemen's Association

www.californiastatehorsemen.com/enviro

Marketing Farm Products

Agricultural Marketing Service

United States Department of Agriculture

www.ams.usda.gov

Institute for Agriculture and Trade Policy

www.iatp.org

Organic Agriculture Centre of Canada
www.organicagcentre.ca

Natural Pest Control

The Barn Swallow: Friend of the Farm
MCE Publications
Maryland Cooperative Extension, University of Maryland
http://extension.umd.edu/publications
> Publishes fact sheet, which provides directions on how to build a barn swallow nest

Bat Conservation International
www.batcon.org

National Pesticide Information Center
http://npic.orst.edu

U.S. Environmental Protection Agency
www.epa.gov/pesticides

Zero Bug Zone
www.zero-bug-zone.com

Recycling

DoSomething.org
http://dosomething.org/recycling
> Recycling information

Earth911
http://earth911.org
> A large database of recycling depots across America. Find recycling centers and chemical drop-off sites and learn how to recycle.

LampRecycle.org
http://lamprecycle.org

SPI: The Society of the Plastics Industry
www.plasticsindustry.org
> The plastics industry trade association which developed the resin identification codes.

Stable/Paddock Management

Hoof Grid
www.hoofgrid.com
> A soil stabilization system developed in Europe

TRAIL-RIDING RESOURCES

American Endurance Ride Conference
www.aerc.org
> American Endurance Ride Conference is a national governing body for long-distance riding.

American Horse Council
http://horsecouncil.org
> The American Horse Council has been promoting and protecting on behalf of all horse-related interests since 1969.

American Trails
www.americantrails.org
> American Trails is the only national, nonprofit organization working on behalf of all trail interests, including hiking, bicycling, mountain biking, horseback riding, water trails, snowshoeing, cross-country skiing, trail motorcycling, ATVs, snowmobiling, and four-wheeling.

Back Country Horsemen of America
www.backcountryhorse.com
> Back Country Horsemen of America is a nonprofit corporation made up of state organizations, affiliates and at large members. We are dedicated to preserving the historical use of recreational stock in the backcountry commensurate with our heritage.

Endurance Net

www.endurance.net

Endurance Net is a resource for those involved in equestrian endurance and distance riding.

The Equestrian Association for the Disabled (TEAD)

www.tead.on.ca

The Equestrian Association for the Disabled works to support the use of trails by disabled equestrians.

Equestrian Land Conservation Resource

www.elcr.org

Equestrian Land Conservation Resource is a national organization dedicated to promoting access to and conservation of land for equestrian use by supporting and facilitating local projects and educational programs for equestrians and conservation personnel.

The Horse Trails and Campgrounds Directory

www.horsetraildirectory.com

The Horse Trails and Campgrounds Directory provides horse, mule, and donkey information about trails, trailheads, horse campgrounds, overnight stay facilities, and bed & breakfasts.

National Recreation Trails (NRT)

American Trails

www.americantrails.org/nationalrecreationtrails

National Recreation Trails program works to preserve and celebrate our nation's pathways.

ENVIRONMENTAL ASSOCIATIONS

Ecological Society of America (ESA)

www.esa.org

Environment Canada

www.ec.gc.ca

Climate change information

Union of Concerned Scientists

www.ucsusa.org

"The Union of Concerned Scientists is the leading science-based nonprofit working for a healthy environment and a safer world. UCS combines independent scientific research and citizen action to develop innovative, practical solutions and to secure responsible changes in government policy, corporate practices, and consumer choices."

INDEX

Page numbers in *italics* indicate illustrations or photographs; those in **bold,** charts.

OTHER STOREY TITLES
YOU WILL ENJOY

Among Wild Horses, by Lynne Pomeranz.
An extraordinary photographic journal of three years in the lives
of the Pryor Mountain Mustangs of Montana and Wyoming.
148 pages. Hardcover with jacket. ISBN 978-1-58017-633-0.

Cherry Hill's Horsekeeping Almanac.
The essential month-by-month guide to establishing good routines
and following natural cycles to be the best horsekeeper you can be.
576 pages. Paper. ISBN 978-1-58017-684-2.

Horsekeeping on a Small Acreage, by Cherry Hill.
A thoroughly updated, full-color edition of the author's best-selling
classic about how to have efficient operations and healthy horses.
320 pages. Paper. ISBN 978-1-58017-535-7.

Naturally Clean Home, by Karyn Siegel-Maier.
More than 100 recipes for effective cleaners that
use safe, familiar, all-natural ingredients.
224 pages. Paper. ISBN 978-1-60342-065-3.

Renovating Barns, Sheds & Outbuildings, by Nick Engler.
Step-by-step advice on how to square and strengthen the structure,
repair or replace the roofing and siding, install new windows and
doors, and even add electricity and plumbing.
256 pages. Paper. ISBN 978-1-58017-216-5.

Stable Smarts, by Heather Smith Thomas.
A treasure-trove of equine know-how, assembled by an
Idaho horsewoman, for riders and owners everywhere.
320 pages. Paper. ISBN 978-1-58017-610-1.

These and other books from Storey Publishing are available
wherever quality books are sold or by calling 1-800-441-5700.
Visit us at *www.storey.com.*